*Royalties from sales of this book will be donated in equal parts
to the charity Bees Abroad and those of
The Worshipful Company of Wax Chandlers of London*

HEATHER HONEY:
A Comprehensive Guide

Michael Badger, MBE

BeeCraft

© 2016 Michael Badger

The moral right of the author has been asserted

All rights reserved. No part of this publication may be reproduced, stored in a retrieval system, or transmitted, in any form or by any means, electronic, mechanical, photocopying, recording, or otherwise, without the prior written permission of the publisher

A catalogue record for this book is available from the British Library

ISBN 978–0–900147–28–9 (hardback)
ISBN 978–0–900147–29–6 (paperback)

Published in Great Britain by
Bee Craft Limited
The National Beekeeping Centre
National Agricultural and Exhibition Centre
Stoneleigh Park
Warwickshire
CV8 2LG

www.bee-craft.com

Cover: Beehives set amongst the ling at Saltergate, North York Moors National Park [William Slinger]

Typeset by Alex Ellis

Photographs are credited individually. The term 'Author's collection' is used in cases where the original photograph is in the Author's possession but the name of the photographer is unknown

Printed in Great Britain by York Publishing Services, York

*'The production of heather honey is quite apart from
the production of other honeys. It requires a special study
and only those who have had a lot of experience with it
know the difficulties.'*

[William Hamilton, Principal Lecturer,
British Beekeepers' Association Coronation Conference,
New Earswick, York, 1953]

(Left to right) Peter Schollick, Gerald Moxon, MBE, Willie Robson and the author, Michael Badger. A visit to the Northumberland fells, August 2014
[Caroline Timms]

Other Bee Craft titles by Michael Badger:
How to use a Horsley Board for Swarm Control, 2009, reprinted 2011, 2012
The Morris Board Method of Queen Rearing, 2011, reprinted 2012, 2015

DEDICATION

*This book is dedicated to my late departed beekeeping mentors:
my father, Tim, Charles S Bell, Arthur Wheildon,
Arthur (Sandy) Powell, Walter Pargeter, Harold J Armitt,
Arthur Horsley, Phillip S Jenkin, Professor Harold Woolhouse,
Colin Weightman, MBE*

*and to all those heathermen, my contemporaries of the past with
whom it has been my pleasure to have made their acquaintance
on the long road of heather-going:*

*the late Stanley Brownridge, Brother Adam, OBE, OSB,
William Hamilton, Bernard Leafe, Dennis Robinson,
Arthur Williamson, Arthur Abbott, Freddy Wilkinson,
Will Snowdin, Arthur Roantree, Dennis Jesper, Bill Reynolds, NDB,
Joe Herrod-Hempsall, Will Slinger, Tom Swaffield,
Harry Allan, NDB, George Green, MBE, Beowulf Cooper,
Harry Grainger, Alexander SC Deans, Alf Hebden, NDB,
Selby Robson, Harry Lockwood, OBE, Fred Richards, NDB,
John 'Eddy' Eade, Harrison Ashforth, NDB, William Clemmit,
Frank Parkinson, MBE, David Batey, Donald Sims, NDB,
Norman Perkins, Thomas Chippendale, Sammy Henderson,
Miss HE Dickinson, Miss Ivy Jacques, Miss Florence Black,
Mrs Susan Holt, Mrs Jean Wood, Haydn Molyneux, Colin Beech,
Bill Bielby, George Leng, Harry Wheldon, Reynold and
Mickey Forshaw, Cecil Tonsley, BEM, Tom and Pat Bradford,
Ted Humphreys, Doug Morris, Bill Lockwood, Lesley Pierson,
Herbert Pierson, Harold Skaiffe, Reg Spruce, Eddie Rix,
John Theobalds, Teddy Sonley, Fred Robson, Walter Salmon,*

*Jack Barker, Geoff Rounce, DFC, NDB, Terry Pearson,
Dr Harold Moody, Martin Phillipson, Charles Cooper,
Sidney Beardmore, Ken Frost, Sir Cedric Aykroyd, Bt,
Ross Thompson, Ken Frankland, Paul Singer, John Read,
George Vickery, Alec Palmer, Bill Parkinson, Alan Hawes,
John Jowett, William (Bill) Foubister, Jim Watson,
Charley Hopwood, Bill Gemmell, Frank Lee, Frank Denton,
Ian Read, Rowland Wood, John Shepherd, Revd Michael Dennison,
Sam Shiptone, Frank Boddy, Mrs Monica Howard,
Peter McScott, MBE, Jim Lawson, Mrs Amy Nicholson,
Maurice Settle, Eric Durham, George Birks, John Stewardson,
Arthur Minnie, Dudley Gue, Don McKay, Jack Renshaw,
Eric Walker, John Foster, John Tyson, J Eric Hughes, Harold Gage,
Bridget Scott, Alan Barber, Harry Bunn, Bill Mackenzie,
Eileen Ramsden, Ted Ramsden and David Marriott.*

*With special mention to the memory of the late Robert Furniss,
New Ridley, Hexham, who did so much with David Pearce,
Phillip Latham and Willie Robson for my dear friend,
Colin Weightman, MBE, in his final years.*

*Finally, to all those heatherman who are present today who
make the annual trek to the heather.*

[*Nota bene*: The generic term 'heatherman' applies equally to either gender. The late Ivy Jacques of Sleights was most proud to be referred to as a heatherman.]

CONTENTS

Acknowledgements 11
Foreword 15
Introduction 17

SECTION 1: HEATHER MOORLAND, ITS ECOLOGY AND MAN, THE BEEKEEPER

A History of How the Heather Moorlands Came About
    to the Present Time 27
From Honey Hunter to Beekeeper 62
The Heather Plants 72
The Heather Moorlands of the British Isles 74
The Heather Honeys 94
Choosing a Moorland Heather Site 102
Section 1 Suggested Further Reading 117

SECTION 2: HEATHER HONEY PRODUCTION

Strains of Bees for Honey Production 123
Types of Hive for Working Heather 150
Systems of Management for Bell Heather
    Honey Production 183
Systems of Management for Ling Heather
    Honey Production 204
Moving Bees to the Moors and Siting Hive Stances 235
Methods Used to Extract and Process Heather Honey 245
The Return from the Heather Moors 290
From Hive to Market: Presentation 298
Ling Heather Honey for Exhibition 303
Final Considerations 311
Section 2 Suggested Further Reading 313

SECTION 3: HISTORICAL ASPECTS OF HEATHER AND OTHER FLORAL HONEY PRODUCTION

A Brief History of Taking Bees to the Heather, Folklore and Little-known Facts 317
Heather Honey for Medicinal Purposes 350

APPENDICES

Appendix A: Fermentation and Hydroxymethyfurfural (HMF) 359
Appendix B: The Dyce Method of Producing Crystallised Honey 364
Appendix C: A Method of Producing Soft-set Honey with Ling Heather Honey 367
Appendix D: Varroa Control for Heather Stocks 372

INDEXES

General 374
People 381
Places 382

ACKNOWLEDGEMENTS

I would like to express my thanks and appreciation to Alex Ellis for his painstaking help and forbearance (that verged on tolerance on many occasions) in the making of this book, for its design and typesetting, illustrations and general advice. Thanks also to Adrian Waring, NDB, for his meticulous review of the text and for the advice that he willingly offered; to Claire Waring for editorial support, general advice on structure and proofreading; to Brian and Margaret Nellist for the excellent photographs that they kindly supplied and for sharing practical expertise and detailed knowledge of the managed heather moorland of north-east Yorkshire; to Bill Cadmore for reviewing the biological details in the text and for his time in explaining issues relating to the genetics of honey bees, biological aspects and the ecology of flowers; to Willie Robson, Chain Bridge Honey Farm, for his time in taking me around the heather moors of the border country of England and Scotland, for his hospitality and for his insight into issues of moorland management; to Tony Harris, Anne Black, Les Webster and Helen Mackenzie for giving up their time to take me around the moors and forests of Aberdeenshire and Sutherland; to Dr David Aston, NDB, for information on honey-bearing heather plants native to the British Isles and their nomenclature, plus the origins of man and his early life on earth; to Dr Sally Bucknall for information on the origins of heather and the ecology of heathlands; to Glyn Davies for sharing his interesting work on queen rearing with mini-nucleus hives, bee breeding and strains of bees; to Dr Dorian Pritchard for reviewing sections on genetics, strains of bees and man as a beekeeper; to Dr Chris Coulson for advice on the botanical aspects of heather and heaths; to William Sutherland, Miriam Rothschild Professor of Conservation Biology, University of Cambridge, for reviewing environmental aspects; to Amanda Anderson, Director of the Moorland Association, for reviewing issues of conservation and historical and modern-day moorland management; to the late Colin Weightman, MBE, for rekindling fond memories of our days together with the late Brother Adam and advice on overcoming production difficulties for ling section heather honey; to Anne Lever for her photographs of Wildboarclough; to Sir John Lister-Kaye, Bt, OBE, for permission to reproduce facts concerning the Scottish heather moors (following his lecture to the Scottish Beekeepers' Association Annual Conference, 2009); to Michael Gleeson, Federation of Irish Beekeepers' Associations, Michael Young, MBE, and Caroline Thomson, Institute of Northern Ireland Beekeepers, for their assistance with information on bell heather honey and the location of heather moors in Ireland; to Gerald Moxon, MBE, for sharing his detailed collection of beekeeping catalogues; to Lester Quayle for his helpfulness in explaining the intricacies of using the Sjolis heather honey loosener; to David Pearce for details of section crates; to David Kemp for snippets of information on working with the late Brother Adam; to Dr Jonathan Munday, MD, Past Master of The Worshipful Company of Wax Chandlers of London, for his comments on the uses of honey for medicinal purposes; to Enid Brown, Ian Craig, MBE, and Norman Walsh, MBE, for their help and assistance with identifying the locations of heather moors throughout Scotland and Ireland; to Ivor Flatman for advice on problems with granulated and soft-set honeys; to Andrew Gibb, David Shannon and Robin Tomlinson for reviews of

Acknowledgements

practical beekeeping; to Graeme de Bracey Marrs, MBE, Past Master of The Worshipful Company of Wax Chandlers of London, for the inspiration to undertake the production of this book; to Mrs Beris Thornton (née Abbott), daughter of Arthur Abbott, for details of her father's early life; to Kenneth S Jackson, former apiarist to Arthur Abbott at Mountain Grey Apiaries, for the background to his former employer; to John Fuller for his comments about the origins of Mountain Grey Apiaries; to Brian Eade for his thoughts and reflections on the heather years from 1950 until the 1990s; to the staff at the East Riding Archives and Local Studies Service, County Hall, Beverley, for assistance with providing historical information; to my fellow directors at Bee Craft Limited for agreeing to publish this treatise on heather honey; finally, to a host of other beekeepers who offered me lots of advice but wish to remain anonymous. It is hoped this book will help and encourage others to become 'heathermen' and enjoy the many challenges in seeking this prince of honeys.

A special thanks for the financial support towards the cost of production of this book provided by Mrs Charlotte Bromet and Nigel Pulling, Yorkshire Agricultural Society, the late Dr Maureen Ziegler and the late Dr Arnold VyVyan-Hobbs.

Finally, my greatest debt is to those around me who supported me throughout the writing of this book: Erica Osborn, who has been a constant source of advice and encouragement, who gave me the basis of many of the ideas, practical suggestions and the continuing willpower over many years to see this book through to fruition; my children, Caroline, Joanne and Andrew, and especially my wife, Hilary, for shouldering the responsibility for family duties so that I could get this book finished. It is to them and my early beekeeping pedagogues that this book is dedicated.

Ling heather cut comb [Brian Nellist]

FOREWORD

*By Graeme de Bracey Marrs, MBE,
Past Master of The Worshipful Company of Wax Chandlers
of London (2010)*

I am delighted to have been asked to write the Foreword to this book on heather honey, its production and uses. Michael's long association with beekeeping and heather honey in particular, both through extensive research and practical experience that extends to over 60 years, should prove to be not only a great help to all those who have an interest in the subject but also an encouragement to go further: to learn, understand and practise the fascinating art of heather honey production.

Graeme de Bracey Marrs, MBE

[Courtesy of the clerk of The Worshipful Company of Wax Chandlers of London]

Heather Honey: A Comprehensive Guide

A wide range of beekeeping gems is offered. Detailed information, supported by many photographs and illustrations, makes this book a pleasure to read. The history, knowledge and instruction captured in this book will aid those who participate in the highly respected British Beekeepers' Association (BBKA) examinations, for which The Worshipful Company of Wax Chandlers of London presents the annual Wax Chandlers' Prize.

The information included about the origins of the landscape of the heather moors portrays a fascinating history. The painstaking reference to all heather areas of the British Isles is exemplary. Examples of man's most recent misdemeanours to the environment are enlightening.

In conclusion, I believe that all beekeepers and others with an interest in this subject and beekeeping as a whole will find this work a fine adjunct to their libraries of books.

Graeme de Bracey Marrs, MBE
July 2016

INTRODUCTION

A couple of years before the death of my father in February 2000, he made a statement (that was recently echoed by my colleague, Peter Schollick), somewhat prophetic at the time, that the art of obtaining ling heather and bell heather honey could so easily fall into oblivion unless those like ourselves gave impetus to new and established beekeepers alike to take honey bee stocks to the moors. He felt this loss would occur fairly soon, quite possibly within the next quarter of a century. Such a prophecy made me feel that I and others, as relatively 'old hands' at beekeeping, have a real duty to do our bit and pass on our knowledge, experience and expertise.

So, inspired by my father and several others, I embarked on this literary adventure over a period of 19 years; a somewhat mythical journey for which I was anything but prepared. The final motivation to get on with it came through some very simple tips from naval historian Dr David Cordingly, former keeper of pictures and head of exhibitions at the National Maritime Museum, Greenwich, and also from Dr David Aston, NDB, president of the British Beekeepers' Association 2014–16, both of whom are no strangers to the ordeals of authorship. Alas, what they failed to mention was that this book's development might well take over my life.

This book is believed to be the most recent detailed textbook devoted to the subject of heather honey production since Dr Stanley B Whitehead's book, *Bees to the Heather*, published 2 April 1954 by Faber and Faber for the princely sum of 12 s 6 d (62.5 p). In addition to the subject of heather honey production, I have included further information that seemed to me to need recording. I felt it would not go amiss to mention the origin of honey bees, man as a keeper of bees, the origins of the upland heather moors, the early history of taking bees to the moors, and some folklore and humour related to beekeeping for heather honey production. I have purposely included this and other additional material to enthuse beekeepers to undertake heather honey production by (hopefully) giving added interest through mention of the ancient, historical and commercial issues that are not found or discussed in current books on bees and beekeeping, other than in the works of the late Dr Eva Crane.

I am not a scientist, nor have I had any training as such, other than a rudimentary knowledge gained from my schoolboy days. Therefore, my interpretation of facts, knowledge and circumstances is written as a layman and as a practical beekeeper.

I am told that I view things both positively and differently from my beekeeping colleagues. To make a subject come alive is my prime aim as I am a strong believer in the truism that enthusiasm can become infectious, especially when it concerns beekeeping.

The intention of this book is to give practical tips to encourage those beekeepers amongst us who may not have visited the moors to participate in this challenging method of honey production. Unfortunately, the majority of books written today, if they mention the subject at all, do so only in a casual manner, leaving the novice in the dark with the feeling that going to the heather is best left alone. This no doubt arises because the authors themselves have no real experience of heather-going. The late William Hamilton's book, *The Art of Beekeeping*, Joseph Tinsley's *Beekeeping Up-to-date* and Alexander SC Deans' *Beekeeping Techniques* give

Introduction

considerable information on heather honey production but are dated in content and style. A more recent book, *Sixty Years With Bees* by Donald Sims, NDB, does devote a chapter to heather honey production. Nevertheless, the content of these works on the subject is useful.

Snags there are with heather honey production, but many are easily overcome, more especially if they are through faulty technique resulting from ignorance of the basic principles. The other difficulties relate to weather and the season, issues over which we have no control. As with gardeners, the beekeeper is the eternal optimist.

There are two types of heather honey available to beekeepers. These are known as *bell heather* and *ling heather*. It has to said that there is a misnomer in the terms used: bell heather is from the genus *Erica*, a heath; ling heather, *Calluna vulgaris,* is regarded as true heather and is the most widespread species throughout the British Isles and mainland Europe. The two heathers can be described simply as types of alpine, low-growing shrubs that have survived for a very long time – through the many ice age periods – within a limited geographical area, generally north of the equator.

The honey from the two types of heather will be dealt with separately as they are two distinct products which are totally different in appearance, colour, texture and other characteristics. They are also obtained by two different methods of honey bee management. Bell heather comes into flower a good month before ling heather, but continues to flourish at the time ling heather comes into flower.

Unfortunately, bell heather is no longer found in vast tracts, as was the case in the pre-Second World War era. The reasons for its demise are not generally understood, yet there

Bell heather in flower in the background with ling heather yet to come into flower in the foreground, Meikle Kinord, Loch Kinord, Aberdeenshire, *circa* 1955 [Edward Jeffree]

are those beekeepers like myself who would work bell heather if the heathland was of a sufficient size to garner a worthwhile surplus. Fortunately, there are still heathland areas throughout the British Isles where meaningful surpluses can be obtained, but they are few in number. They are: the Mountains of Mourne, Northern Ireland; Winfrith Heath, Dorset; for some years at Cannock Chase, Staffordshire; Muir of Dinnet, Aberdeenshire; and Thorne Waste, East Yorkshire.

Bell heather honey and ling heather honey, it has to be said, are rarely easy crops to obtain in quantity owing to a number of unique issues. The available areas of bell heather are not as extensive as those of ling heather. Ling requires special preparations to be made in order for a meaningful surplus to be achieved. The main consideration is the weather that is encountered on the upland moors. It is often unfavourable

Introduction

and something the beekeeper has no control over. This gives rise to a short, often intermittent, honey flow over no more than a three-week period at best. In addition, colder temperatures are experienced at moorland altitudes where the ling flourishes. While the bees will continue to forage for nectar from other sources after this period, if the ling honey crop is left on the hive, a blended honey will be the end result.

At the end of July, honey bee colonies are losing strength; it is instinctive for colonies to reduce their populations in response to the shortening day length. Therefore, it is difficult to maintain the strength of colonies in preparation for the moors without resorting to special management measures and techniques. The narrative gives detailed procedures to ensure the beekeeper can maximise the outcome of his or her efforts to the full.

The long distances colonies may need to be transported can make going to the heather questionable as a financial proposition, because of the cost of fuel, unless a number of stocks are taken. However, a quantity of this honey is worth all the work and effort involved. On the plus side, the advantages of modern-day off-road vehicles and the excellent road networks make sites more accessible to those who live great distances from the heather moors. Stocks rarely return without a reasonably provisioned brood box of natural stores of honey and pollen.

Looking back to my own formative years of 'heather-going', just as the motorway age was dawning, I would travel to the area of Wildboarclough (on the county border between Derbyshire and Cheshire) with my mentor, the late Harold J Armitt, from his home at Sarehole Road, Hall Green, Birmingham. I remember enjoying cups of coffee with cold bacon sandwiches at daybreak, after bringing our stocks along the wild and lonely roads leading to the heather moors. Oh, happy days.

A decade later, my trips to Timble, Denton, West End and Dallowgill moors in Yorkshire, with Peter Hardy and the late Phil Jenkin, invariably ended with us enjoying a pint or three with other like-minded heathermen – Robin Tomlinson and sadly deceased beekeepers Alan Hawes, Harry Grainger and Terry Pearson – who also needed to quench their thirsts by heading to the Timble Inn, overlooking the beautiful Washburn Valley and Fewston and Swinsty reservoirs.

In the early 1970s, Will Slinger, Bernard Leafe, Colin Weightman, Dennis Robinson, Martin Phillipson, Walter Salmon, J Eric Hughes, George Leng, Tom Bradford, William Clemmitt, Charley Hopwood, Frank Denton, Dennis Jesper,

The author on the heather moor, Leek, Staffordshire, 1961 [Charles Bell]

Introduction

Bill Lockwood, Bill Reynolds, Maurice Settle and Arthur Williamson – sadly all deceased – would gather each year at the Yorkshire Beekeepers' Association Honey Show, De Grey Rooms, York, to discuss the merits of the exhibits of heather honey. They would review the judges' awards (since they themselves were judges as well as exhibitors) and were gracious in winning as in defeat. Listening to these sages of beekeeping, these were, indeed, the balmy days of beekeeping for me. Now there is the present generation who are too numerous to mention by name; I am bound to miss out a loyal comrade. That would not do.

It needs to be recorded that the writing of this book, which first saw the light of day in 1997, was the whim of my late father, Jack 'Tim' Badger. I was encouraged to complete it by Graeme de Bracey Marrs, MBE. As a book of a specialist nature, its development became fraught with issues of pragmatism and readability in relation to how best to present the subject matter in a way which makes it flow for the reader so that he or she may most easily digest the complex matter of practical heather honey production. In drawing up the topics for discussion and presentation, it became apparent to me that wider aspects of general beekeeping would have to be discussed and included as a means of getting to the core subject. I make no apology for extending my brief to much broader horizons. My colleague, Robin Tomlinson, complimented me on the full scope and detailed narrative relating to both the original Demaree method and its current adaptations.

I am also conscious that I have drawn on and written extensively about people who are no longer alive and who will be unknown to the majority of current-day beekeepers. It is these people who passed on to me their consummate skills, knowledge and wisdom related to heather-going which, in turn, I am pleased to pass on to a greater audience. I accept that I have fallen into the age-old trap of our elders: 'As we grow old we tend to live on our memories'. Having kept bees since the time a king was on the throne, it seemed an essential requirement that I, too, pass on my own bits of practical beekeeping knowledge. The historical facts come as part of my armoury and if not recorded by me now will be lost forever. This is particularly true in relation to my recording of the historical facts concerning the origins of the 'MG' – now 'Modified National' – hive. Fortunately, some reference to its origins is recorded in ROB Manley's book, *Beekeeping in Britain*, first published in 1948, at a time when the British Beekeepers' Association was making moves to work with the British Standards Institution to agree a 'standard' Modified National hive.

I hope the critics of my work will take this summation into account when this book is reviewed.

Michael Badger, 17 April 2016

SECTION 1:

HEATHER MOORLAND, ITS ECOLOGY AND MAN, THE BEEKEEPER

An ideal accessible moorland site within the North York Moors National Park that is sheltered from westerly winds, has water close by, and both bell heather and ling heather in close proximity [Michael Badger]

A HISTORY OF HOW THE HEATHER MOORLANDS CAME ABOUT TO THE PRESENT TIME

Nature reclaims what man neglects (Anon)

To understand the origins of the heather moorland areas as we see and know them today requires reference to the beginnings of the British Isles and the formation of these islands. Key stages of their development can be summarised as follows:

Cryogenian period	c720,000,000–635,000,000 BP
Pliocene epoch	c5,300,000–2,600,000 BP
Pleistocene epoch	c2,600,000–11,700 BP
last interglacial (Ipswichian)	c130,000–115,000 BP
last glaciation (Weichselian)	c115,000–11,700 BP
Holocene epoch	c11,700 BP to present
Palaeolithic	to c10,000 BC
Mesolithic	c10,000–4,500 BC
Neolithic	c4,500–2,500 BC
Bronze Age	c2,500–800 BC
Iron Age	c800 BC–AD 43
Roman	cAD 43–410
Anglo-Saxons and Vikings	cAD 410–1066
Normans	cAD 1066–1154
Middle Ages	cAD 1154–1536
Post-medieval	cAD 1536 onwards

In simplistic terms, the British Isles is a consequence of the ice ages that have carved it out over the millenia coupled with man's intervention in the landscape at certain points in history.

DEVELOPING MOORLANDS

The heather moorlands as we see them today came about as a result of two distinct periods in our history. Firstly, man's intervention in the present interglacial period with the repeated burning of grassland areas and, secondly, in the past 300 years, through the establishment of driven grouse shooting estates and hefted sheep farming operations in upland moorland locations.

The most recent period of intervention in moorland locations came about as a result of the actions of Queen Victoria and her consort Prince Albert in the 1850s, following the purchase of the Balmoral Estate. The transformation of this estate was to lay the foundations of the well-managed moorlands that we know today. Sir John Lister-Kaye, Aigas Field Centre, Beauly, Inverness, spoke on this subject at the Scottish Beekeepers' Annual Conference in September 2009:

> *Scotland has made a highly favourable impression on us both. The country is really very beautiful, although severe and grand, perfect for sports of all kinds ...*
> [Prince Albert's diaries, 1842]

It is well documented and recorded that, at least a century before, the Duke of Sutherland's notorious Highland

How the Heather Moorlands Came About

clearances (that took advantage of lax tenurial law and were organised by his agents) saw the introduction of sheep farming into moorland areas. This started the control of the rough grassland that dominated the landscape. The vast numbers of sheep kept down the grass which, in turn, allowed heather to grow and flourish unhindered.

The Highland clearances ended the lifestyle of the indigenous crofters, a way of life that they had followed for centuries. These proud and industrious people were driven from the land, the majority of them emigrating to the Empire to seek a new life.

The vast areas of blanket bog on highland moorland soon put paid to the quick profit expected from sheep farming; the capability of sheep to live off such a landscape had been much over-rated. The moorland soon returned to the wild as sheep numbers declined.

Other major landowners throughout the UK, seeing the transformation of the landscape, were quick to follow by caring for moorland areas in very similar ways, but without deposing the tenants of their holdings. The types of sheep introduced to these moorland areas were breeds that readily endured the harsh climate. They could survive on both marginal grassland and on the heather itself.

The land agents began to realise that some form of land management was necessary. The simplest way to rid the moorland of straggly heather bushes was with fire. It has to be said that in the early 1800s, moor burning was haphazard, casual and sporadic. With the introduction of carefully monitored and controlled burning of the moors in the 1850s, straggly heather plants were kept in check, enabling them to produce a continuous supply of young shoots upon which both sheep and moorland grouse for driven shooting interests were able to feed.

It is accepted that upland heather moorland is a semi-natural habitat that has been managed and mismanaged over the centuries. Upland moorland is not a natural environment, unlike the rainforest, for example, which can and does sustain itself. Both ling heather and bell heather require looking after and managing. If not managed correctly, the heather, though a low-growing shrub, grows into a dense mass of long woody stems that, after time, supports very little wildlife. Such heather has no grazing or economic value and grows to the point that it is very hard to walk through.

The Moorland Association states that Britain has 75 per cent of the world's entire resource of open heather moorland. Following the Second World War until the mid 1990s, some 81,000 ha (200,000 acres), 20 per cent of the heather moorlands in England and Wales, were lost to overgrazing, afforestation, the spread of bracken and general neglect.

The Moorland Association was formed in 1986 to halt the loss of heather moorland and to secure its future. Members are responsible for 348,000 ha (860,000 acres) of the estimated 389,000 ha (950,000 acres) of heather moorland remaining in England and Wales, albeit for driven grouse shooting interests. It is because of the upland moorland habitat produced by past management that 60 per cent of the heather moorlands in England and Wales have been designated as Sites of Special Scientific Interest (SSSIs), Special Protection Areas or Special Areas of Conservation. It has to be said that it is not generally understood by the majority of the public that heather moorland is a fragile

How the Heather Moorlands Came About

Ling heather at Wildboarclough [Anne Lever]

ecosystem and its bird and small mammal populations are easily reduced by too many predators.

It is no coincidence that most of the actively managed heather moorlands in the British Isles are in private ownership, since this has provided the necessary continuity of management to preserve them effectively.

CONCERNS FOR MOORLAND MANAGEMENT IN THE FUTURE

There are industries and livelihoods that are affected by the management of heather moorland. The management policies of newly introduced government agencies are being questioned by those living on and around the upland moors who see their way of life being affected. These people include shepherds, gamekeepers, farmers and the like, whose expertise is of a specialist nature. These are regarded by many as the true custodians of moorland areas. Their expertise in moorland management has been handed down by generation after generation of moorland keepers.

Since the 1950s, environmentalists have expressed concerns relating to the draining of damp moorland to expand the areas of drier, driven grouse moorland, later to be managed by burning of the heather. Some environmentalists believe that no one really knows what an original natural heather moor looks like as burnings greatly affect the moorland flora.

The breeding of both red and black grouse relies on the availability of young heather for the birds to feed upon. Managed burning is the moorkeepers' way of ensuring regeneration of heather plants and new growth which, in turn, enables grouse to prosper. The environmentalists' view is that those moorlands which are unburnt, or infrequently burnt, have a tendency to have more varied flora; the upland moorland is no longer a Victorian playground, solely for driven grouse shooting interests. Burning, or 'swailing' as it is known in Devon, is regarded by some as a practice upon which there is over-reliance.

Conflicting views will, no doubt, continue. Government handouts to landowners may exacerbate the situation, with financial incentives to leave moorland areas to nature.

Nevertheless, modern-day heather moorland is a man-made construct that is now valued as a unique habitat that, left to nature, would disappear fairly quickly. It may not be a Victorian playground anymore, but without management the moors as they are enjoyed now would soon be no more. A succession of vegetation from heather, through bracken, birch woodland, scrub oak to mature forest would occur.

MANAGED HEATHER BURNING

Heather burning needs to be undertaken early in the year, before 10 April or, exceptionally, before 25 April. The dates differ in England, Scotland and Wales. Heather burning after these dates is regarded as irresponsible. It deprives ground-nesting birds of cover if carried out in May or later.

Burning can give bracken regrowth a headstart over that of heather. It is a major concern that the deep rhizomes of bracken are unaffected by the heat generated and it is the first plant to show recovery after moor burning. Despite the high cost, the use of helicopters to spray large areas of moorland to remove bracken has been resorted to in locations where bracken has been allowed

Bracken spraying [Brian Nellist]

Burning of the heather moor

[Courtesy Amanda Anderson, the Moorland Association]

Damping down the fire without the use of water

[Courtesy Amanda Anderson, the Moorland Association]

Husband and wife undertaking managed heather burning on Egton Moor, north-east Yorkshire [Author's collection]

to dominate through the absence of active management by moorland keepers. The banishment of bracken is well favoured by sheep farmers as it naturally harbours sheep ticks (*Ixodes ricinus*). Increased prevalence of this parasite soon arises as bracken management slips.

If the heather becomes too old before it is burnt, there is no regeneration of the plants. The natural lifespan of ling heather plants is 30 years at most. Careful heather burning has a positive effect on heather seeds. It seems to encourage them to sprout into life, a fact often overlooked by those who have little or no knowledge of the process. If the ground is too dry, however, the heat may be so severe that it destroys the seeds and they are no longer able to geminate to create new plants.

Sheep and grouse are good companions on the moor. The sheep eat the young heather, which ensures that the tender regrowth is available for the grouse. This, in turn, is when the beekeeper can also benefit from the nectar-bearing flowers on young plant growth.

MAN AT HIS WORST

Man's actions in the two world wars were most shameful occurrences. A typical example is seen in the areas of hardwood and native pine trees in the Scottish Highlands of Strathspey and Rothiemurchus, Inverness-shire. Large woodland vistas were felled on an unimaginable scale to provide timber for ammunition boxes. The old pines of Lochiel Old Forest, in the same locality, were destroyed by deliberately setting fire to forest areas for commando training in both conflicts. Sadly, neither of these two large areas of desecrated woodland were replanted until many years later. 'Sustainability', as we know it today, was unheard of. No doubt,

Deforestation in the Second World War, Sutherland, 1943 [F Fraser Darling]

Acres of moorland affected by a fire near Whitby in 1976 [Brian Nellist]

The same moorland, near Whitby, 25 years after the fire [Brian Nellist]

war department officials played on landowners' patriotic duty as it is understood that they were barely compensated, receiving only minimum payment for these priceless commodities, let alone any consideration for future replanting.

STAKEHOLDER MANAGEMENT ISSUES

The lessons of history require the various stakeholders of the landscape to manage moorland burning with greater care. Restricting the use of burning as a management tool leads to unforseen consequences and conflict. The moors as we know them today are to be managed for the benefit of future generations. Fortunately in 1988, Nigel Lawson, Chancellor of the Exchequer, put an end to the tax breaks for the forestry industry, thus saving the remaining open expanses from being planted with conifers. Interestingly, in 2015, the new specialist fund-management group, Gresham House, announced that it had taken an interest in vast swathes of Scottish woodland, mainly because it does not incur inheritance tax and, apparently, over the past ten years has produced annualised returns of 18.8 per cent compared with 6.6 per cent from UK stocks and shares. The investment management team says: 'The attraction of investing in trees is that they grow, irrespective of whether financial markets are rising or falling. They grow not only in volume but also become more valuable as they mature'. The IPD UK Annual Forestry Index has outperformed both equities and bonds over the past 20 years (November 2015).

A recent example has arisen that illustrates the problems. Concern relating to the management of bird life on upland moorland reached new heights in summer 2014 with the

How the Heather Moorlands Came About

issue of illegal killing of hen harriers, which are known to predate on red and black grouse. All those involved have tried to reach a consensus over future management to resolve the growing conflict between driven grouse moor custodians with commercial interests and bird conservation groups.

Black and red grouse are unique among game birds in that they cannot be reared successfully in captivity as the adults require a sole diet of ling heather. To maintain or increase their population can only be undertaken by careful management on the fell. Predation from hen harriers is exacerbated because these birds live in colonies and because one brood needs hundreds of meals.

The proposed plan is to accept one breeding pair of hen harriers in an area of four square miles. If numbers rose above this level, eggs would be removed and taken to an incubation centre. The reared chicks would be released into areas of Devon, Cornwall and Wales, away from the driven grouse moors, thereby re-establishing these species elsewhere. This plan by the Department for Environment, Food and Rural Affairs (Defra) has not met with uniform approval from conservation groups.

The Moorland Association represents the owners of most of the 149 driven grouse moors of England and Wales. Its members are responsible for over 348,000 ha (860,000 acres) of moorland, preserving 1500 jobs. Grouse shooting is worth £67.7 million a year to England's rural economy.

Over the past 25 years, grouse moor owners have regenerated and recovered 217,000 acres of moorland, an area larger than Birmingham. They have blocked 1250 miles of drainage ditches as an aid to locking up more carbon, creating thousands of small ponds and scrapes that will conserve water voles, amphibians, insects and pearl mussels in watercourses by trapping sediment and slowing down run-off. This will also reduce potential flood risks below the moor edge.

Forestry Enterprise (England) stated in November 2015 that a review of Sitka spruce (*Picea sitchensis*) forests, planted in the 1970s, is to be undertaken. The management of these forests, especially those that crowd the edges of streams, rivers and watercourses, has had a devastating effect on the freshwater pearl mussel (*Margaritifera margaritifera*) that plays an important part in biodiversity. This indigenous mussel lives for over 100 years, at the bottom of fast-flowing streams. Illegal fishing also affects these long-lived invertebrates. Any work required in a watercourse that contains pearl mussels is a criminal offence if the mussels are damaged in the process.

THE GROUSE SPECIES THAT INHABIT THE UPLAND MOORLANDS OF THE UNITED KINGDOM

Whether it is accepted or not by the heather-going beekeeper, he or she derives benefit from the well-managed upland moorland of privately owned and managed estates. These moorland areas, with their large swathes of both ling heather (*Calluna vulgaris*) and, to a lesser extent, bell heather (*Erica cinerea*), provide the food source for the birds of driven grouse shooting interests.

Upland moorland habitats are man made; left to nature, they quickly becomes strewn with bracken, acid grassland, mattgrass, sheep's fescue, straggly heather and bilberry bushes, which will not sustain a large population of these

How the Heather Moorlands Came About

species of game birds. Although young grouse chicks feed on insects and young heather at an early age, their sole source of sustenance as mature birds is young ling heather, supplemented with grit. Other game birds, such as pheasants and partridge, can be reared by artificial means as their diets can be supplemented with corn and mixed meal with enhanced vitamins.

Ling heather grows in the northern parts of the UK, the Isle of Man and the Republic of Ireland, and certainly prospers most in areas of driven grouse shooting. It is the management of the upland moorland for driven grouse that has given rise to the beautiful scenery that is the mainstay of the national parks.

Worldwide, there are many related groups of grouse. Four species are native to the UK, three of which are endangered owing to particular circumstances, some of which are understood and others not so. The known factors leading to endangerment and, ultimately, extinction are human activity, changing climate, intensive agriculture, deforestation and increased industrial activity.

The Grouse Birds of the British Isles

The four types of grouse common to the UK are:

- the red grouse (*Lagopus lagopus scoticus*) – a medium-sized bird regarded by some authorities as a subspecies of the willow grouse. The red grouse is the bird reared for driven grouse shooting
- the black grouse (*Tetrao tetrix*) – a larger bird that is a protected endangered species found mainly in the Pennine chain and throughout Scotland, but not in Ireland

- the ptarmigan (*Lagopus mutus*) – a medium-sized bird slightly larger than a grey partridge that is also a protected endangered species. It is found only in the Scottish Highlands. It breeds high up in the snow-covered mountains, seldom moving far from its breeding sites. There are other species commonly known as the willow ptarmigan, rock ptarmigan and the white-tailed ptarmigan. All of these are found in the northern hemisphere but rarely in the UK
- the capercaillie (*Tetrao urogallus*) – a huge woodland grouse; the large black males are unmistakable. They spend a lot of time feeding on the ground but may also be found in pine trees, feeding on shoots. There are localised species found in Scottish native pine wood (a rare and vulnerable habitat) and in commercial conifer plantations. The UK capercaillie population has declined so rapidly that it is at very real risk of extinction (for the second time in living memory) and the bird is a 'Red List' species.

The true ancestor of all grouse species is the wild turkey (*Meleagris gallopavo*). Grouse derive from the New World, one to three million years ago, when they occupied the boreal forests of the uppermost northern hemisphere in North America, Europe and Asia. The boreal forests form a broad circumpolar band immediately south of the Arctic Circle in a vast expanse that easily rivals the rainforest regions of the world. This band runs through most of Canada, Russia, the northern British Isles and Scandinavia. The northern boreal eco-region accounts for about one-third of the world's total forest area.

How the Heather Moorlands Came About

Red grouse at Wildboarclough [Anne Lever]

Various species of grouse evolved as a result of geographical isolation caused by changes in climate over time. Grouse are adapted to a cold environment, without need for migrating southwards as other birds and animals do. The ability to survive on a sparse diet of coarse food sufficient to sustain them through long cold winters makes them matchless in evolutionary terms for adaptation to cruel harsh environs.

The Politics of Grouse Shooting

It is worth a mention as to how these vast areas of upland grouse moors came into the ownership of current landowners. By chance, William the Conqueror won the crown of England through trial by battle at Hastings in 1066 and the death of King Harold on the battlefield. (Trial by battle was a procedure which had been brought to Britain by the Normans following the conquest; it was not present in Saxon law.) He awarded himself all the land in the kingdom. Those closest to him by patronage were allowed to keep their land holdings. Unlike elsewhere in Europe, the UK has never succumbed to a revolution. Thereby ownership remained in the hands of the descendants of those benefiting at that time.

Over the centuries, there have been upheavals in land ownership from the Dissolution of the Monasteries and the Industrial Revolution to the present day with foreign financiers and city bankers buying up and acquiring the estates of the British aristocracy. Such an example is the 14,000 ha (35,000 acre) Gunnerside Estate, North Yorkshire, sold in 1995 by the Earl of Peel to Robert Warren Miller, the American billionaire tycoon of the Duty Free Shoppers chain.

It should be acknowledged that a greater part of society abhors any sport or interest that involves the killing of a wild animal for sport or social activity as this is regarded as cruelty. Cruelty to animals comes in many forms. It is beyond the scope of this narrative to discuss in any depth the morality of hunting wild animals, other than to say that everyone should respect lawful activity and, in doing so, educate the public – particularly the young – with a sense of moral responsibility towards living creatures. To its credit, the driven grouse shooting fraternity has a strict code of shooting that ensures participants are experienced gunmen, thereby minimising unnecessary suffering through wounding as opposed to outright kill. Well-trained gun dogs pick up all shot birds.

The view presented by many protest groups is that driven grouse shooting carries an unwelcome history of money, power and politics that is deeply rooted in the British class system and the establishment. They contend that such

How the Heather Moorlands Came About

shooting is a leisure activity reserved solely for the rich and powerful, seemingly overlooking the ethics related to the killing of a wild creature. Driven grouse shooting is an expensive leisure pursuit with the cost amounting to £36,000 per day (in 2015) for participation in the top driven grouse shoots of the UK.

Recently, it has been stated that rural communities benefit substantially from grouse shooting, year on year. A parish survey around Blanchland, Northumberland, found that 55 per cent of hotels and guest houses directly or indirectly involved in driven grouse shooting increased guest numbers during the four-month shooting season, pushing up the hotels' annual average occupancy rate from 50 to 65 per cent. The driven grouse shooting industry continues to play a big role in helping to maintain, restore and preserve the local landscape. Grouse is also served in many local pubs, hotels and restaurants.

So what is the attraction of an expensive day at a driven grouse shoot on upland moorland? Generally speaking, for ardent shooting enthusiasts, it requires considerable skill to shoot and kill a flying bird that is both fast and furious, travelling at speeds in excess of 125–140 mph. In shooting terms, there is nothing to rival or compare with the experience. Experienced driven grouse shooters are excellent practitioners whose consummate skill, speed of reaction and accuracy are tested to the limits when shooting a red grouse in flight. It can truthfully be said, irrespective of what is thought of driven grouse shooting as a sport, that it is a uniquely British experience, practised nowhere else in the world, deep rooted in history and culture for a very small minority who, in the main, are the landowners and the very rich. That said, the shoots themselves are supported by people from all walks of life, including those working as 'beaters', dog handlers and in a host of other roles on which the activity relies.

The Rise of Driven Grouse Shooting from 1850 to 2016

The introduction of the breech-loading shotgun in the 1850s coincided with the interest shown by Queen Victoria and Prince Albert in the rural pursuits of deer stalking and driven grouse shooting. Prior to 1850, grouse shooting in the Scottish Highlands was undertaken with the use of specially trained dogs. Moorland management for grouse shooting was in its infancy, with managed burning yet to become a widespread practice. This form of grouse shooting was known as 'dogging'. It required specially trained dogs, which used both sight and scent.

Historians of driven grouse shooting recognise four key periods. The first is the period from the early 1840s to the middle of the First World War. The second is the inter-war period that saw a sharp decline in activity, then a recovery, followed by another decline prior to the outbreak of the Second World War. The third period saw a general recovery to a peak in the 1970s, followed by decline into the mid 1980s. There has been a resurgence of popularity in the fourth period of recent years. It is believed there are 460 active driven grouse moors in the UK of which 147 are in England.

In the past decade, a vocal minority is seeking public support for driven grouse shooting to be abolished. The Royal Society for the Protection of Birds (RSPB) is pursuing

How the Heather Moorlands Came About

a policy of grouse moor licensing as a means for government involvement. No doubt once legislation is passed, single-action minority groups will put pressure to bear on compliant politicians to become more actively involved, ensuring that amendments would be made to the current Animal Welfare Act 2006 and Animal Health and Welfare (Scotland) Act 2006. This may, in turn, lead to the eventual abolishment of driven grouse shooting.

From a personal perspective and with rural interests, I am sickened by all forms of animal cruelty. Involving central government in moorland affairs is far from the way forward; proper stakeholder management is a much better way. Excuse the use of a cliché, but as my great uncle would say, 'You get more out of tickling than scratching'.

ENDING OF THE ICE AGE

This last ice age is believed to have ended around 11,700 years ago, ushering in the beginning of the Holocene epoch that still persists today. The end of the Younger Dryas period (often referred to as the 'big freeze') saw the return of rising sea levels, caused by melting glaciers, that in all probability cut Britain off from Ireland and continental Europe, creating the land mass and coastline as we know it today. The return of warmer climatic conditions gave rise to a different type of woodland landscape: the creation of hardwood forests of oak, ash, sycamore, horse chestnut, beech and lime. The change of climate also saw the arrival of a number of different types of animal that required different hunting techniques from those used previously.

This new landscape contained woodland clearings and extensive open spaces that attracted wild animals to feed on the lush grassland, enabling hunters to stalk and kill them more easily, with the use of trained hunting dogs.

The training of dogs enabled stone age man to round up wild deer, goats, boar and aurochs (ancient cattle) and, through selective breeding, it was possible to introduce a level of domestication to livestock. Tended by minders (shepherds), animals foraged on open plains and in woodland areas during the day. At night, they were kept in fenced enclosures within fortified compounds as a means of protection from other raiding animals.

From these early homesteads came the first farmers. Using trained dogs to help manage livestock thus ensured food was available on a regular basis. This enabled these people to discard a nomadic lifestyle and adopt a more stable existence. A more consistent food supply supported an increasing population. This settled lifestyle was the beginning for man to organise into family groups.

THE GENESIS OF HEATHER MOORLAND

Recent archaeological research (through pollen analysis, deoxyribonucleic acid [DNA] testing and excavation) has found that that there were extensive and widespread reed-bed margins around these woodland areas and that they were regularly set ablaze in summer, purposefully by man and sometimes by lightning. These burnings, whether deliberate or by nature, encouraged vigorous regrowth of young grassland because of the potash deposits left by the ferocious fires that devoured all before them, very much like areas of

Africa that today are frequently ablaze as a result of natural lightning strikes. These early people soon latched on to the method of using fire for the regeneration of grassland. Rain quickly reinvigorated the soil and new grass appeared for animals to feed upon. Burning was the only method open to these early settlers to ensure a continuous supply of new grass for livestock.

To this day, grassland is of vital importance for livestock raised for human consumption and for milk and dairy products. It is possible these early people developed some sort of 'winter provender' for their retained livestock, as it would be in their interests to overwinter the animals to provide them with milk from both the goats and aurochs. Wild deer would be hunted down using dogs and killed when needed.

LONG TERM EFFECTS

Regular burnings were to continue from 8,500 BC for about 8,000 years and were to prove, in time, to have devastating consequences for the future landscape and for later peoples. Upland moorland burnings often ran out of control (which they do to this day) and are believed by modern-day ecologists to have led to the creation of thin fragile soils that, over time, led to soil erosion and allowed heath, heather and bracken to flourish. Heaths and heather plants readily adapt to these ecological conditions and there are few, if any, nutrients to support other vegetation. This largely lifeless upland was soon passed over by man and left to nature until the arrival of the Normans and, from about AD 1080, the arrival of the monks from France.

ROMAN TIMES AD 43–410

The arrival of the Romans in AD 43 saw the introduction of new forms of livestock including sheep, rabbits, pigs, pheasants and possibly honey bees, too. The Romans are believed to be the originators of the term 'an army marches on its stomach'. The Romans were renowned for being town dwellers, building large communal conurbations like Bath, York, Lincoln and London. They created a network of roads to link up the various communities and to support trade with country folk at fairs and markets. Researchers and historians have established that beeswax was used as a tradable commodity in Roman times when it was used for the seals of documents, writing tablets, and the waterproofing of leather.

ARRIVAL OF THE ANGLES, SAXONS, JUTES AND VIKINGS AD 410–1066

The departure of the Romans in AD 410 saw the established Roman towns fall quickly into disrepair during the following decades as the invading Angles, Saxons and Jutes lived a simple countryside lifestyle, similar to the original inhabitants of Britain. Angles, Saxons and Jutes were Germanic people. They had some organised farming methods with land being divided into farms of 50–200 acres. The Angles and Saxons originated from Saxony and colonised areas of southern England.

 The arrival of the Vikings in AD 793 did little to improve the landscape or ecology of the countryside, other than that they did introduce some order into farming settlements with the introduction of an early form of enclosure system.

How the Heather Moorlands Came About

We know that these people were beekeepers of some merit; the earliest British archaeological remains of skep apiculture come from the Anglo-Norse town of Jorvik, now modern York. As an aside, skep beekeeping has been researched in Wilhelmshaven, Germany, from the remains of a woven wickerwork skep dated 200–0 BC.

NORMANS AD 1066–1154

The arrival of the Normans and their investment in the land saw a new way of living for inhabitants of the British Isles. *Circa* AD 1080, monks and other monastic orders arrived from France. They were renowned as stockmen and horticulturists, bringing with them large flocks of sheep that were soon to suppress the more vigorous vegetation in the grassland of upland areas. It is inevitable that monasteries in moorland locations would have harvested both heather honey and heather-based beeswax.

The Domesday Book included many entries relating to the ownership of swarms, honey and wax obtained from honey hunting in the forests throughout England. Forest beekeeping continued in England up to the 1880s. Straw skep beekeeping persisted until the mid 1880s, although it was still in evidence with some beekeepers until the 1930s and with one or two in North Yorkshire in the 1960s.

THE MIDDLE AGES

The end of the Norman period saw the rise of the Plantagenet dynasty with increasing prosperity, despite the channelling of wealth to finance wars and the various crusades. During this

A collection of straw skeps at Hartpury College, Gloucestershire, 1961
[Author's collection]

A skep on a skep stone at Battleton Farm, Kineton, Warwickshire, 1953
[Author's collection]

How the Heather Moorlands Came About

time, beeswax was used as a payment mechanism for taxes owed to the king, with certain officials having the right to demand a given number of candles and pounds of beeswax.

It is well documented that religious ceremonies used large quantities of beeswax; this was especially true in France. Among the annual revenues of the Bishop of Puy was 20 lb of beeswax. In 1330, the farmers of the domain of Beauregard each had to pay 2 lb of beeswax annually. In 1632, John de Frettar, sexton of the monastery of Chaise Dieu, stipulated that the tenant was to bring to his house a rental of 600 lb of beeswax each year on St John's Day, this to be of good merchantable quality. Another deed, dated 27 July 1668, shows that for six years, John Marel rented the workroom at the monastery for the payment of 120 lb of beeswax, to be paid in candles of first quality.

During this period there was a trend for the use of log hives throughout Europe, although this never caught on in Britain, with most apiarists persisting with skep hives until the mid 1880s when the first cottager hives and then the WBC hive were introduced.

The use of skeps generally involved the sacrifice and destruction of the bees, though the practice of 'drumming' to entice the bees into a new skep was also used. Having removed the bees by one method or another, the beekeeper would cut out the combs containing only honey, then combs containing brood would be removed and, finally, any remaining odds and ends of wax. Honey was extracted from the combs by placing them in a cloth bag and allowing them to drain. Then more honey of lesser quality was removed by wringing the bag and its contents. Finally, the crushed combs, the raided skep and the cloth bag would be steeped or gently heated in water to dissolve out any honey. After straining, this liquid was used as the basis for the production of mead.

Swarm control with skeps was a very simple operation that involved inverting the skep, placing it on a special board to stabilise it and adding a roof-type board that retained the heat and also allowed the bees to fly. The upturned position of the skep caused the bees to pull down any queen cells quickly as the orientation was then unnatural. Within a few days, the skep was returned to its upright position. The skeppist repeated this procedure each time he observed queen cells. My great uncle, who kept a number of skeps, told me that having inverted the skep more than twice the bees would be ready and waiting for the beekeeper in no uncertain terms.

THE WORSHIPFUL COMPANY OF WAX CHANDLERS OF LONDON

The use of beeswax candles, which up to that time was chiefly confined to churches, monasteries and the houses of the nobility and landed gentry, became general in the fifteenth century as it was the only means of lighting. The trade of candle making acquired so much importance that the guild of wax chandlers of London obtained an act of incorporation. This guild is now known as The Worshipful Company of Wax Chandlers of London. The lower classes used candles made of tallow, the refined fat of cattle and other livestock. Tallow also became controlled by a guild of tallow chandlers.

The Wax Chandlers company gained ordinances in 1371, with a petition presented to the Court of Alderman. In that year, it would appear that the guild had attained

The Wax Chandlers' hall, Gresham Street, London [Michael Badger]

an established position. In 1484 (the year after which Richard III was crowned, whose body was found in 2013 beneath a municipal car park in Leicester), the company was incorporated by royal charter, with powers to choose a master and two wardens to oversee the craft of wax chandlery and, upon any defects being found, to punish the offenders. This charter was confirmed and later extended by subsequent charters.

DECLINE IN THE USE OF BEESWAX IN MEDIEVAL ENGLAND

The Dissolution of the Monasteries Act, 1535, was the beginning of the decline in the use of beeswax in religious houses. It was estimated that 372 houses in England and 27 in Wales were closed down and their assets seized by the crown. The seizure of wealth was at its peak in 1541, the year in which the Dissolution of the Monasteries ended, with Thomas Cromwell being executed as a heretic on 28 July 1540.

Besides the reformation, which reduced the demand for beeswax, other products appeared in competition such as stearin, paraffin and ceresin, which are waxes obtained from various animal and mineral sources. These waxes were used frequently for adulterating beeswax, thus lowering its price.

FROM SKEPS TO MOVABLE-FRAME HIVES

The life of skeps would have been very short if they were left to the elements. For 600 years or more, from AD 1300 to AD 1900, specially designed bee boles were

South-facing bee bole on the side of a barn on the Pickering road, north of Old Malton, North Yorkshire [Michael Badger]

perfected to provide outward protection to straw skeps from rain, sun and frost, to prevent rotting of this natural material. These bee boles were of many types and variants and included simple wooden and stone structures. The International Bee Research Association (IBRA) commissioned a register of bee boles in 1952 that identified over 618 such sites throughout Great Britain and Ireland.

Beekeeping in the mid 1800s saw the introduction of movable-frame hives. The watershed between fixed comb beekeeping and movable-frame hives came about from 1851, following definition of the bee space by the Revd Langstroth in the United States of America (USA). It would be another 30 years before the movable-frame hive came into general use in Great Britain. These hives had the colloquial name of 'cottager hives' for those used by the working classes and, soon after, the WBC hive, named after its

57

A bee shelter at Kirkby Stephen, North Yorkshire [Author's collection]

Movable-frame hives alongside skeps under a bee shelter, Scargill, County Durham [Author's collection]

designer William Broughton Carr, for the wealthier classes. During Victorian times, there was even class distinction with beehives.

The introduction of movable frames saw the beginning of modern beekeeping and the decline of the use of straw skeps. The blind naturalist, François Huber, had perfected a leaf hive in 1750, whereby the combs were arranged like the leaves of a book. As early as 1682, movable-frame hives had been written about by Sir George Wheler in *A Journey into Greece,* following his visit to the hives at St Cyriacus's monastery on Mount Hymettus, Attica, Greece.

The collection of honey and beeswax by the rural working classes continued and, no doubt, the inhabitants of the large houses collected honey for their own consumption from hives within their gardens. The lower classes would have gained a small income from the locally produced honey and wax. Those near the moors would have benefited from a crop of heather honey.

TO MORE RECENT TIMES

From around 1563, the introduction of cane sugar in quantity from West Indian sugar plantations reduced the demand for honey. Cane sugar had been available as a sweetener from the Far East since the 1300s but was a very expensive commodity. The gentry used sugar while the masses continued to use honey. Fortunately, the wholesomeness of honey could never be tainted with the issues of slavery.

The Great Beekeeping Exhibition at the Crystal Palace in 1874 saw the formation of the British Beekeepers' Association, with Welsh, Scottish and Irish associations following after.

The founders of the British Beekeepers' Association were vainly hoping that the other countries of the British Isles would join them, hence the term British rather than English.

Very soon, the early pioneers of beekeeping in the United Kingdom came to prominence, such as Thomas Cowan and William Broughton Carr. There was also a number of several beekeeping publications that, sadly, are now defunct. These included the *British Bee Journal* and the *Beekeepers' Record*. The brothers William and Joseph Herrod-Hempsall were the main figures of British beekeeping of the early twentieth century, up to 1919. William Herrod-Hempsall was appointed as the Ministry of Agriculture's Adviser in Beekeeping. In this exalted position, he had considerable influence that many said he abused. His autocratic manner saw resentment from a number of progressive county beekeeping groups. The Kent Beekeepers' Association, through Jim Wadey, founded *Bee Craft* as an alternative voice and was of the view that beekeeping had a wider focus. The need for change to the British Beekeepers' Association came to a head in 1943 with the introduction of a new system of management that exists along similar lines to this day.

The 1920s saw the rise of eminent commercial beekeepers: Manley, Gale, Madoc, the Ratcliffe Brothers, Abbott and Buckfast Abbey, with a number of other large-scale operators conducting business locally within each county.

A demand for honey and beeswax has continued throughout the twentieth century. In the case of beeswax, demand reached its highest level in the 1960s because of its use in the manufacture of munitions for the Vietnam War (1955–1975). Beeswax is the perfect medium to use as a water/damp inhibitor for both high explosives and

Tonnes of beeswax, Snainton Bee Farm, Scarborough, North Yorkshire
[Les Chirnside]

small arms ammunition. The Vietnam War saw the price of beeswax rise to unbelievable levels as the US govenment purchased every tonne available on the world market.

On a more peaceful basis, beeswax was used as a coating agent of well-known sweets consumed by young and old alike, but it is now replaced by an alternative product.

FROM HONEY HUNTER TO BEEKEEPER

The availability of honey for sustenance and consumption came first to the bees themselves, followed by those animals that trod the earth centuries before early man came into existence. These animals would have been those which either lived in the trees, or those which lived on the ground but had the ability to climb trees to raid the honey bee nests in the cavities in tree trunks. The use of simple rope made from wild honeysuckle (*Lonicera* spp.) gave early man the resource to make simple ladders to climb trees or cliff faces. It begs the question: which came first, the ladder or the wheel?

A wild honey bee colony in the burr of a sycamore tree [Alan Woodward]

THE FIRST FLOWERS

According to the most recent research, it seems that the first fossilised flowers came from about 130 million years ago. There are some expert archaeologists who believe that flowers may have been around 250 million years ago, although fossils that old have yet to be found.

There are other experts who maintain that flowers arrived through several evolutionary phases that occurred in relatively quick succession. In evolutionary terms, this means a span of two to three million years or longer.

Irrespective of the timing, we have to accept that flowering plants and trees have made a colossal impact on life on earth. There is a general consensus amongst experts that without them, life would be very different as 70–80 per cent of all food eaten by humans comes from flowering plants and fruit-bearing trees. A key issue is whether flowers were first insect-pollinated or wind-pollinated.

Flowers are the defining characteristic of angiosperm plants that enable reproduction and spread. These plants dominate our world illustrating the success of their reproductive strategy. Evolution was at its most creative when the first flowers evolved. Clearly the colourful or scented flower evolved in harmony with the evolution of insects. Although flowers may have pre-dated pollinating insects by some 30 million years, there is general agreement between botanists and entomologists that flowers and pollinating insects evolved together in a process called 'co-evolution'. It could be argued that the two developed ways and means of helping each other survive and, therefore, succeeded through a symbiotic relationship that embraced mutual co-operation

A relatively young forager bee loaded with ling pollen on an alighting board
[Brian Nellist]

and coordination. One provided the transportation and the other provided food. Another theory is that flowers appeared, insects visited them, flowers improved, insects improved and an evolutionary spiral took off.

Through the evolutionary process of time, extending over many millenia, flowers have developed and produced a great number of inducements to get the various animal species on land and in the air to distribute their pollen and seeds. Flowers need pollinators as much as pollinators need flowers.

THE ARRIVAL OF BEES FROM A WASP-LIKE ANCESTRY

The bees themselves are thought to have evolved from wasp-like insects about 150 million years ago. These insects are believed to have developed a preference for nectar and pollen over the carnivorous diet of other insects. This transformation from carnivore to herbivore may have arisen from the lack of continuous availability of meat-type food because of natural competition with other carnivorous insects. The ancestral bee found an alternative readily available food supply in nectar and pollen from trees and flowers. Whether by design or fault is not readily understood other than that the evolutionary process stepped in at some stage.

The honey bee is classified thus:

Kingdom: animal – the animal kingdom is the highest and most comprehensive, defined by cellular structure.

Sub-kingdom: Annulosa – one of the primary groups of animals with bodies enclosed in a rigid external skeleton.

Phylum: Arthropoda – a sub-division of Annulosa, animals with jointed feet.

Class: Insecta – a class of articulate, usually winged animals, with three pairs of legs and bodies divided into three distinct segments: head, thorax and abdomen.

Order: Hymenoptera – an order of insects with four membranous wings.

Family: Apidae – a group that comprises bees of many kinds including bumblebees, solitary bees and honey bees.

Genus: *Apis* – a group with one common feature that involves storing food reserves by various but differing

	methods. This group includes: *Apis dorsata*: giant honey bee, found in Asia; *Apis florea*: little honey bee, found in Asia; *Apis cerana indica:* found in Asia and the Indian sub-continent; *Apis scutellata*: found in Africa; *Apis mellifera*: western honey bee, found throughout the world.
Species:	*Apis mellifera* – a group that bears an almost exact likeness with possible minor differences (eg, Italian, Caucasian, etc). These can interbreed to produce hybrids of varying vigour and suitability. In the United Kingdom (UK) it seems that it would be impossible to have a pure-bred species due to the mongrelisation of the indigenous bees. Resorting to artificial insemination and its fallibilities would fail in all aspects. Most bees in the UK are subspecies as a result of importation of bees from abroad.

MAN AS A BEEKEEPER

Initially, man was a 'hunter-gatherer'. He was not a practising beekeeper, in the true sense of the word, using proper equipment. The time when a more sophisticated approach was adopted is unknown. Suffice to say that bees as we know them in their present evolutionary form are believed to have come into existence around 55–65 million years BC. Early

From Honey Hunter to Beekeeper

humans would have been quick to home in on the bounty of honey, no doubt through the observations of bears and other animals stealing it from wild bees' nests.

It is accepted that the beginning of the Holocene epoch, 11,700 years ago, saw the global spread of early man, *Homo sapiens sapiens,* who would have been quick to exploit the great forests and cliff escarpments, home to many wild colonies of honey bees. These early humans latched on quickly to harvesting honey and marked trees to indicate sources. We know from the rock paintings found in India and South Africa that early man may not have been a 'true' beekeeper, but as a honey hunter he undertook an organised activity in teams. These early peoples were fearful of nothing; a few stings would be nothing to the dangers they would encounter regularly in the toil of daily life.

THE USE OF SMOKE

In time, man came to realise that bees were easily calmed by the use of smoke. But why? The reasons are not generally understood. Fires were regular occurrences throughout the large forests, caused by lightning strikes or by man. It is thought that feral bee colonies were driven from their nests by forest fires. The bees, sensing the smoke, instinctively filled their honey crops in readiness to decamp to a new location. An emergency migration would require the same stocking up on food as at the time of swarming.

The effects of smoke are also believed to suppress guard bees' senses while undertaking guard duty; fewer foragers will leave the nest. The effects reduce the bees' natural defensiveness in response to a threat. There is an unproven theory that smoke muffles bees' alarm pheromones during disturbance to the colony.

Early man learned to use smoke as an ideal instrument when plundering wild nests for honeycomb. Why get stung unnecessarily if smoke acts to calm the bees while their nests are raided?

HONEY HUNTERS

Until recently, the earliest known record of man obtaining honey from bees was a Mesolithic painting of honey hunters on the wall of a rock shelter at La Arana, eastern Spain, dated *circa* 6000 BC.

Mesolithic man was often competing with other predatory animals for tree and cliff honey. At some stage, believed to be

Honey-collecting scene, *circa* 6000 BC. Rock painting in La Arana shelter, Bicorp, eastern Spain [Copy by E Hernandez-Pacheco. Enlarged detail]

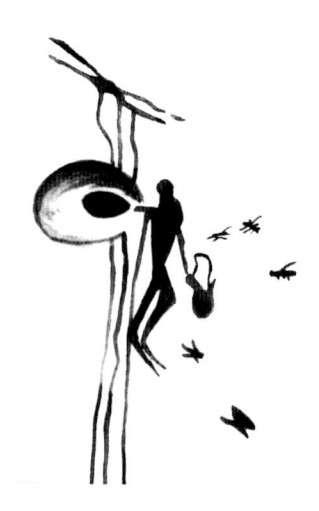

Honey hunting in Nepal [Claire Waring]

from 5500 BC, man decided to locate trees that housed feral colonies for his own use. These trees were cut down, chopped into manageable log hives of about 1.5 m in length and hauled to a specially constructed bear-proof compound adjoining the main living compound. No doubt man had several of these tree stump hives for his use. These hives were later adapted so that the beekeeper could access the honey by a trap-door closure at the rear. One could say that man had become a beekeeper. Nonetheless, honey hunters in Nepal undertake the beekeeping traditions of Mesolithic man to this day.

EARLY REFERENCES TO BEES

Use of honey and wax for commercial purposes more or less followed after early man had sourced these products from nests in trees and on rock faces. It may well be that these two commodities were used as a means of barter between hunter-gatherers.

There are references to honey in several passages of the Bible. Honey and wax were known to have been commercial entities in 4000–500 BC among the ancient civilisations of the Sumerians who settled on the banks of the Tigris and the Euphrates. This area is often referred to as the fertile crescent, Babylonia or, in modern geography, as Mesopotamia. The source of the Euphrates is in eastern Anatolia; Anatolian bees, no doubt, originated from this region.

Around 3400 BC, the Egyptians used honey bees as a sign of the king. Much documentary and pictorial evidence of bees, honey and wax was found in the tombs and mausoleums, as in the tomb of Rekhmire, 1450 BC, at Thebes, Upper Egypt (on the West Bank at Luxor). A painted relief shows jars of honey being filled by attendants with combs being taken from hives with the aid of smoke. There are fragmented remains of a painting showing hives at the sun temple at Sakkara, Cairo (2400 BC). It is believed that these are the earliest known records of beekeeping and harvesting honey. The Egyptian colonists who settled in Greece, Cyprus, Crete and Palestine took bees with them and bees that are found in these areas today still have the traits of bees found in Egypt.

THE HEATHER PLANTS

There are two types of plant from which different honeys, both known as heather honey, originate.

- Ling heather. True heather, *Calluna vulgaris*. The common name is derived from the old Norse *Lyng*.
- Bell heather. Five related ericaceous species of heath. Some species are unique to the locality where they are grown. *Erica cinerea* is found throughout Great Britain. There is also the cross-leaved heath, *Erica tetralix*, the Cornish heath, *Erica vagans*, the Dorset or Arn heath, *Erica ciliaris*, and the Mediterranean heath, *Erica carnea*.

Calluna vulgaris, *Erica tetralix* and *Erica cinerea*, Fylingdales, North Yorkshire [Brian Nellist]

It is not so long ago (in living memory) that there were abundant areas of bell heather throughout the British Isles at Cannock Chase, Staffordshire; the Dinnet Valley, Aberdeenshire; Thorne Waste, East Yorkshire; Esk Valley, North Yorkshire; Bradfield Moor, South Yorkshire; Blanchland and Rothbury, Northumberland; Stanhope, County Durham; Winfrith Heath, Dorset; the Dorset Heights; and the Mountains of Mourne, Northern Ireland. Each of these particular sources gives an indivually flavoured honey as a result of local soil and environmental conditions which can, and do, vary between districts.

Ling heather, along with the cross-leaved heath, *Erica tetralix*, is found extensively on the northern moorland uplands which extend for many miles. These locations are found with few, if any, trees because of the high altitudes and exposed windswept surroundings. *Erica* is found extensively in Hanoverian Germany, at Lüneburger Heide, and in the Picos de Europa mountains, Spain.

Ling is widespread throughout the world and is found from the northern parts of the mountains of north Africa to the Arctic Circle, including westward in Greenland and Newfoundland, and to the east in the Ural Mountains of Siberia.

THE HEATHER MOORLANDS OF THE BRITISH ISLES

The heather moorlands of the British Isles cover about 25 per cent of British uplands (15,000 km^2), with about half of this area managed for driven grouse shooting interests. The moorland habitat is varied. Three broad but distinct types can be recognised and identified: heathland with bell heather (*Erica cinerea*), cross-leaved heath (*Erica tetralix*), ling heather (*Calluna vulgaris*) and bilberry (*Vaccinium myrtillus*); mire or bog made up of a variety of mosses, sedges and small shrubs; acid grassland with sheep's fescue, wavy hair grass and mattgrass. Most, if not all, of these upland sites are each designated as a Site of Special Scientific Interest (SSSI).

The moor at Wildboarclough. Note the reddish hue of the ling [Anne Lever]

Ling heather is an alpine plant that has been in existence for a very long time. It is the sole member of its genus. It probably came about through mutations in another member of the Ericaceae family, almost certainly in an isolated population, so that, in time, it differed sufficiently to be considered distinct. It may then have spread so that it now co-occurs with the plant from which it originally evolved. It could be argued that, in the main, the plant's characteristics developed as a result of particular conditions relating to climate, geography and soil. It is difficult to be specific about where the plant originated other than to say that it lies within western Europe.

Heathland may be present in localities where conditions are relatively cool and humid. In the northern hempishere, these are usually sub-arctic high-altitude mountainous areas. Such areas are characterised by an absence of trees.

It is without doubt that both climate and soil influence the presence or absence of certain types of vegetation. The annual rainfall in heathland areas of the British Isles is 60–110 cm (23.5–43.5 in), well distributed throughout the year. Spring and autumn are somewhat long, summers are occasionally very hot and winters are often mild. Such a climate is termed a cool temperature oceanic system, whereby the available water supply exceeds losses by evaporation.

Although the effects of climate change may well alter matters in the next few centuries, since 1975 the British Isles has experienced long periods of drought and excessive rainfall. It is a fair assumption that the British Isles and areas of north-west Europe have a unique environment, ideal for growth of heather and heath.

THE ECOLOGY OF HEATHLAND AND MOORLAND

Heath is a word that is believed to have its origins at the onset of Anglo-Saxon times and possibly earlier. The term refers by definition to under-shrub vegetation consisting of, or dominated by, members of the Ericaceae family of heaths and heathers. Furze and broom are other under-shrubs. This group of plants differs from other shrubs as the plants are permanently low growing, cannot develop into trees and have a lifespan of no more than 30 years.

It can be said that lowland heathland is vegetation that has replaced forest over time; this may or may not be true or correct for every site. The loss or disappearance of forests is, in the main, usually the result of human action and intervention so, to a large extent, heathland can be regarded as a habitat created and maintained by human activities. It needs to be noted that a careful balance is required to maintain such habitat, otherwise its loss can be, and often is, extensive.

Many legends are attached to the term *moor* (or *muir* as is used in the Scottish Highlands). While it is fashionable today to use the term to describe uncultivated hill-land, such as Ilkley Moor or the Lammermuir Hills, moors are normally characterised by a layer of peat varying in depth from a few millimetres to many metres thick.

Scientists, ecologists and conservationists often disagree whether there is a real distinction between heath and moor. An example is the 'downs' of the Lizard peninsula, Cornwall. This area is referred to as heath, but excavations by researchers have found that it has, or has had, a cover of

Leaching podzol, Morayshire [Michael Badger]

peat. Heaths are found for the most part in the drier parts of the UK. They experience recurring periods of drought, whereas moorland areas are located in areas of high rainfall.

The majority of heaths are found on acidic soils. An interesting phenomenon is that ling is known to alter the soil characteristics to produce a soil type known by its Russian name of *podzol*. The situation which leads to podzol formation arises through plant debris that accumulates on the surface of the land and rots down to form a type of leaf mould. The top layer of mineral soil immediately below this leaf mould is noticeably bleached to a white or pale grey colour. At a further depth of 150 to 300 mm, the soil forms a hard black layer that is referred to by environmentalists as a *humus pan*. Below this is a much harder concrete-like rust-coloured layer known as an *iron pan*. These different pans result from the percolation of rainwater which penetrates the leaf mould, dissolving the organic acids present.

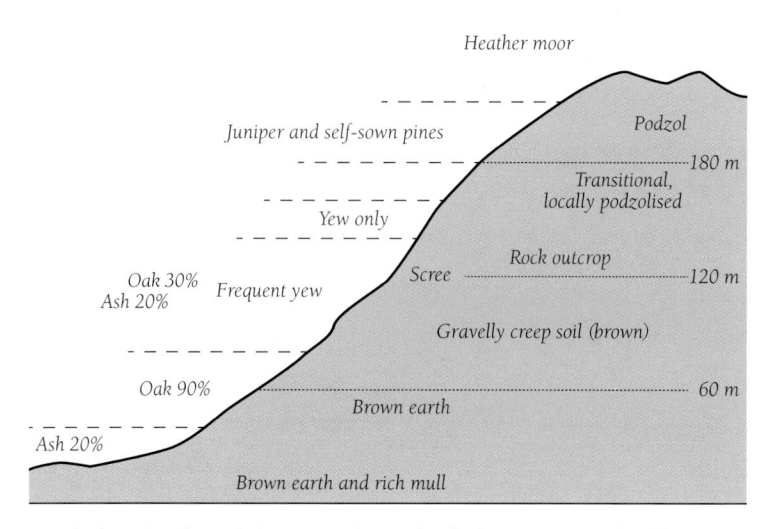

```
                                    Heather moor
                                  _____
                               /              \
              Juniper and self-sown pines    Podzol
              _ _ _ _ _ _ _/                           ────180 m
                                      Transitional,
              _ _ _ _ _ _ _           locally podzolised
                    Yew only
              _ _ _ _ _ _/
                                Rock outcrop
                         Scree  ──────────────────────120 m
    Oak 30%   Frequent yew
    Ash 20%
              _ _ _ _ _ _/     Gravelly creep soil (brown)

              Oak 90%          ...............................60 m
                               Brown earth
    _ _ _ _ _/
    Ash 20%
              Brown earth and rich mull
```

A typical section through the vegetation and soil of an upper moorland location

Leaching of humus and iron compounds present in the topsoil occurs which has the effect of leaving it a lighter colour. These substances are re-deposited at lower profiles, giving rise to cementation of the soil particles and thus creating the pan conditions. The formation of podzol requires the soil to be sufficiently acidic and certain chemical compounds to be present in the decaying leaf mould. These compounds are found in ling heather. Once formed, podzol is known to persist for well over a million years.

Podzolic soils are widespread on the drier moorlands. This peat is formed from the remnants of centuries upon centuries of dead plants. In areas where the ground is more or less permanently wet, the vegetation does not ultimately rot down but piles up on the surface, forming a black or dark brown layer that is quite distinct from the mineral soil below. Therefore, peat is not the same as the leaf mould found in dry areas of heath and woodland. A location where the peat layer is many metres thick and has water running through it constantly is known as 'blanket bog'. The Flow Country blanket bog extends from the tip of Caithness, Scotland, and continues south through Sutherland for approximately 95 miles. It is regarded by many as one of Europe's natural wonders because of its vastness, covering approximately 4,000 km^2. Hopefully, in time, this location will become a Unesco World Heritage Site.

Blanket peat can be used as a fuel. It is cut from man-made gullies often referred to as hags. The remains of plants and vegetation, preserved over a million years of the peat's formation, can be seen in the exposed vertical sides of the gullies.

Peat workings in Meenavaghran, County Donegal, March 2014 [Erica Osborn]

Ling heather, Bradfield Moor, Sheffield; an extensive moor that is rarely burnt but is managed with both sheep and cattle [Ivor Flatman]

Bell heather and cross-leaved heath, Fylingdales, North Yorkshire [Michael Badger]

A rare species of bell heather, Tomintoul, Banffshire [Michael Badger]

MOORLAND LOCATIONS

The most notable moorlands of the British Isles include the Dark Peak, the Pennine chain from Derbyshire to Northumberland, north-east Yorkshire and Cleveland, the Forest of Bowland, the Lake District, mid Wales, the Lowlands and Southern Uplands of Scotland, the Scottish Highlands, a few very small pockets in western Herefordshire, Staffordshire, East Yorkshire and isolated areas of Northern Ireland and the Republic of Ireland. The Isle of Man also has a number of notable areas.

The heather moors used by beekeepers are shown on the maps of the British Isles (following) and described in the key below.

Note. The text refers to the original county areas rather than the political area names set up under the 1974 revision to the county/metropolitan areas.

Heather Honey: A Comprehensive Guide

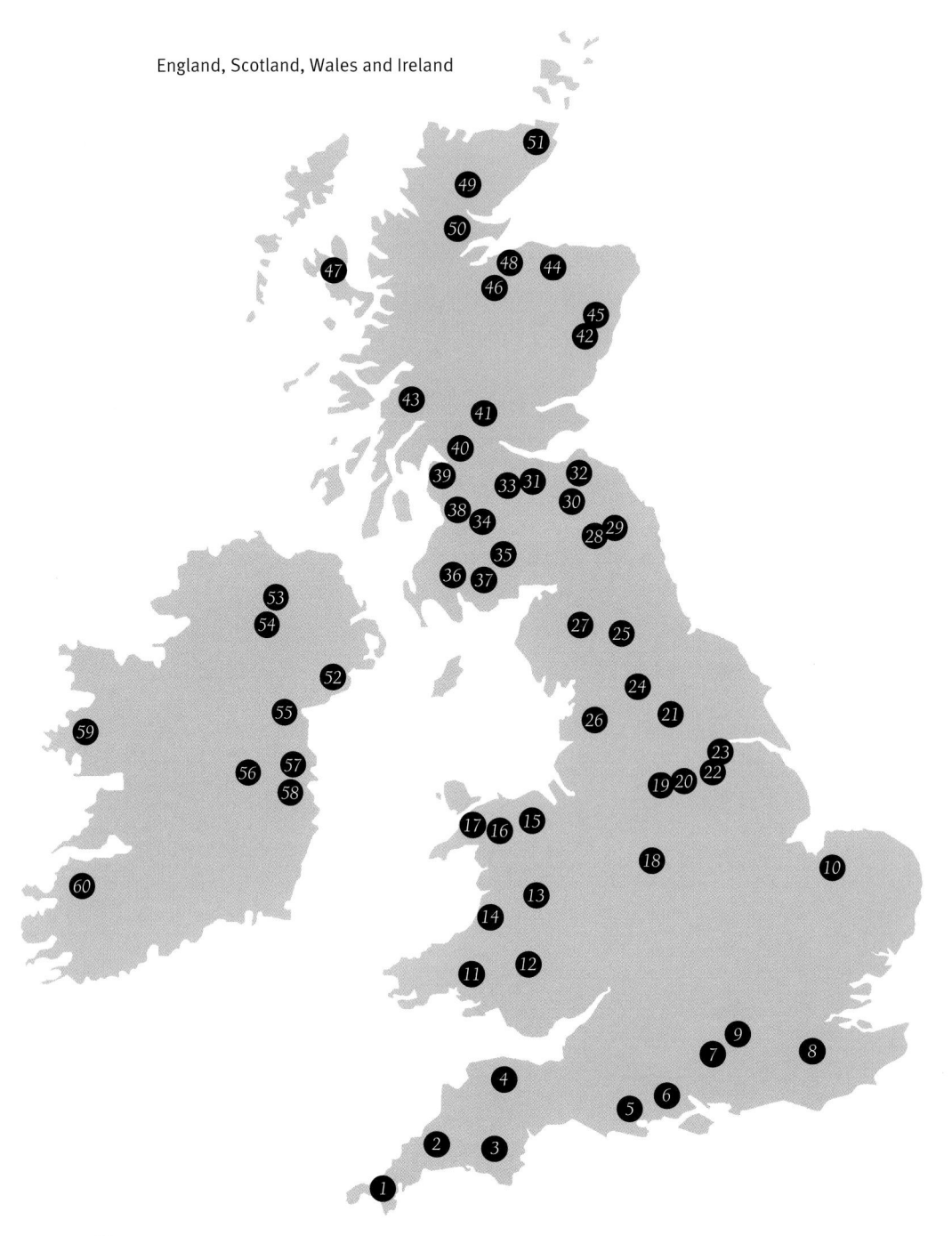

England, Scotland, Wales and Ireland

1. South-west Cornwall. Mainly heaths on the Lizard peninsula and north of Penzance with small patches of *Erica cinerea*, Cornish heath, cross-leaved heath and other species.
2. Bodmin Moor. Ling and small patches of *Erica cinerea*, Cornish heath, cross-leaved heath and other species.
3. Dartmoor. Ling and small areas of *Erica cinerea* with cross-leaved heath and other species.
4. Exmoor. Ling and small areas of *Erica cinerea*, cross-leaved heath and the rarer Dorset heath, *Erica ciliaris,* are found with other species.
5. Dorset Heights. Small areas of *Erica cinerea* and cross-leaved heath and the rarer Dorset heath, *Erica ciliaris,* are found with other species.
6. Hampshire. Heathland in the New Forest where *Erica cinerea* and cross-leaved heath are extensive.
7. North-east Hampshire. Ling and small areas of *Erica cinerea* with cross-leaved heath are common.
8. Sussex. Small heath areas in and around the north Sussex Weald.
9. Surrey. Bagshot Heath and surrounding heathlands extending westwards to the Hampshire border. Ling and small areas of *Erica cinerea* with cross-leaved heath are common.
10. Norfolk. Small heath areas to the east of King's Lynn.

11. Carmarthenshire. Ling and small patches of *Erica cinerea* and cross-leaved heath with other species.
12. Brecknockshire. Parts of the Black Mountains and Brecon Beacons. Ling and small patches of *Erica cinerea* and cross-leaved heath with other species.
13. Radnorshire. Ling and *Erica cinerea* throughout the moorlands and slopes of the Welsh mountains with cross-leaved heath and other species.
14. Cardiganshire. Ling and *Erica cinerea* throughout the moorlands and slopes of the Welsh mountains with cross-leaved heath and other species.
15. Denbighshire. Ling and *Erica cinerea* throughout the moorlands and slopes of the Welsh mountains with cross-leaved heath and other species.
16. Merionethshire. Ling and *Erica cinerea* throughout the moorlands and slopes of the Welsh mountains with cross-leaved heath and other species.
17. Caernarvonshire. Ling and *Erica cinerea* throughout the moorlands and slopes of the Welsh mountains with cross-leaved heath and other species.
18. Cannock Chase, Staffordshire. Relatively poor heathlands with *Erica cinerea*, cross-leaved heath and other species.
19. Derbyshire. Bleaklow, Dark Peak and above Buxton to the Cheshire border at Leek, Wildboarclough. Heather moors east of the River Derwent. Ling heather with small areas of *Erica cinerea*, cross-leaved heath and other species.

20. South Yorkshire. The Pennine chain including Bradfield Moor running up to the borders of Derbyshire, Cheshire, West Yorkshire and Greater Manchester. Good ling heather predominates with small areas of *Erica cinerea* and cross-leaved heath with other species.
21. West Yorkshire. The Pennine chain including good heather moorlands at Emley, Ilkley, Timble, Haworth and Rombalds. Moorland running up to the borders of Lancashire, Derbyshire, Cheshire, North Yorkshire, South Yorkshire and Greater Manchester. Good ling heather predominates. Small areas of *Erica cinerea* and cross-leaved heath with other species.
22. Thorne Waste, East Yorkshire. Relatively poor heathlands with patches of ling, *Erica cinerea* and cross-leaved heath with other species.
23. Crowle and Hatfield Moors, East Yorkshire. Relatively poor heathlands with patches of ling, *Erica cinerea* and cross-leaved heath with other species.
24. North Yorkshire. The Pennine chain including good heather moorlands running up to the borders of Lancashire, Cumbria, Derbyshire, Cheshire, East Yorkshire, West Yorkshire and Greater Manchester. Also West End, Harrogate, Skipton to the Aire Gap, Ripon, Dallowgill, Richmond, Catterick, Middleton-in-Teesdale, Settle and Appleby-in-Westmorland. Good ling heather predominates. Small areas of *Erica cinerea*, cross-leaved heath and other species.

25. County Durham. The Pennine chain including good heather moorlands running up to the borders of Lancashire, Cumbria, North Yorkshire, East Yorkshire, Tyne and Wear and Northumberland. Also Scargill, Reeth, Barnard Castle, Stanhope, Stainmore and Brough. Good ling heather predominates. Small areas of *Erica cinerea*, cross-leaved heath and other species.
26. Lancashire. The Pennine chain including good heather moorlands running up to the borders of Cumbria, North Yorkshire, West Yorkshire and Northumberland. Also Saddleworth, Oswaldtwistle, Rossendale, Ingleborough, Forest of Bowland, Darwen, Haslingden, Rivington and slopes of the Eden Valley. Good ling heather predominates. Small areas of *Erica cinerea* and cross-leaved heath with other species.
27. Cumberland. The Pennine chain including good heather moorlands running up to the borders of Lancashire, Yorkshire and Northumberland. Also the Lakeland fells and slopes of the Eden Valley. Good ling heather and small areas of *Erica cinerea*, cross-leaved heath and other species.
28. Northumberland. The Pennine chain including good heather moorlands running up to the borders of Lancashire, County Durham, Tyne and Wear. Also the Scottish Borders and eastern slopes of the Tyne Valley. Good ling heather and small areas of *Erica cinerea*, cross-leaved heath and other species.
29. Roxburghshire. Cheviot Hills and termination of the Pennine chain including good heather

The Heather Moorlands of the British Isles

moorlands running up to the borders of England, the hills south of Jedburgh to the southern slopes of Northumberland. Good ling heather and small areas of *Erica cinerea*, cross-leaved heath and other species.

30. Selkirkshire. Good ling heather with small areas of *Erica cinerea* are to be found around the southern slopes of the Tweedsmuir Hills to the Moffat Hills and in and around the Ettrick Forest. Throughout the forest, the areas of heather have declined over the past 60 years.

31. The Moorfoot Hills, Lothian. These belong to the Southern Uplands that stretch from Peebles northwards towards the south of Edinburgh. The moors are predominately ling heather found mainly in the central valleys but over the years this has declined because of the invasion of bracken and other species.

32. Lammermuir Hills, East Lothian. These extend to Lothian and southwards to the borders of England. Ling heather is predominant, especially on the north and south slopes.

33. Pentland Hills, Lothian. These stretch from south-west Edinburgh to beyond Biggar and to Greater Clydesdale. Ling heather and *Erica cinerea* are confined to the southern slopes.

34. Lanarkshire. Excellent ling heather with small areas of *Erica cinerea*, cross-leaved heath and other species found in the area of the Leadhills, Tinto Hills and Culter Fell to Crawford.

35. Dumfries and Galloway. Langholm, Moffat, moors at Moniavie, Mennock Pass, Thornhill and north-east of Gretna Green. Good ling heather predominates with areas of *Erica cinerea*, cross-leaved heath and other species.
36. Wigtownshire, Dumfries and Galloway. North of Newton Stewart and areas extending throughout the Galloway peninsula and north and south-west from Kirkcowan. Good ling heather predominates with areas of *Erica cinerea*, cross-leaved heath and other species.
37. Kirkcudbrightshire. Laurieston Moors north of Castle Douglas and between Newton Stewart and New Galloway. Good ling heather predominates with areas of *Erica cinerea*, cross-leaved heath and other species.
38. Ayrshire. Notable areas include West Kilbride to Fairlie in the north of Ayrshire to Darvel and Cumnock. Good ling heather predominates with areas of *Erica cinerea*, cross-leaved heath and other species.
39. Renfrewshire. Around Muirsheil Regional Park. Excellent ling heather, small areas of *Erica cinerea* and cross-leaved heath.
40. Dunbartonshire. Excellent ling heather with small areas of *Erica cinerea* and cross-leaved heath from the Clyde to Loch Lomond, north and east of Garelochhead.
41. Perthshire. Several moors around Aberfoyle, Loch Earn, Comrie, Crianlarich, Tyndrum, Luib, Strathyre and Crief. Good ling heather

predominates with areas of *Erica cinerea* and cross-leaved heath.

42. Kincardinshire. The hills are very good for ling heather with small areas of *Erica cinerea*, cross-leaved heath and other species.
43. Argyll and Bute. Very good for ling heather with small areas of *Erica cinerea* and cross-leaved heath in and around Oban, Kilmartin, Loch Eck, Ardentinny, the Kintyre peninsula and islands of Jura, Islay and Mull.
44. Banffshire. Very good for ling heather with small areas of *Erica cinerea* and cross-leaved heath.
45. Aberdeenshire. The largest moorland in the UK. Very good for ling heather. Areas of *Erica cinerea* growing freely with cross-leaved heath in the Dinnet Valley, Cambus O'May and Kennethmont.
46. Inverness-shire. Very good for ling heather. Areas of *Erica cinerea* growing freely with cross-leaved heath in Tomatin, Strathnairn, Daviot, Glen Truim, Drummossie and Glen Moriston.
47. Skye and Raasay. Ling heather grows throughout the islands.
48. Moray and Nairn. Ling heather grows very well throughout the Spey, Lossie and Findhorn valleys, with areas of *Erica cinerea*.
49. Sutherland. Very good for ling heather around Lairg, Brora, Loch More and Melvich, with areas of *Erica cinerea* growing freely and cross-leaved heath.
50. Ross-shire. Very good for ling heather with areas of *Erica cinerea* growing freely and cross-leaved heath

in areas of Kildermorie, Strathrory, Tain Hill and stretches from Braemore to Garve.
51. Caithness. Ling heather areas patchy throughout Dunnet Head, Brough, Scotscalder, Altnabreac and Latheron. Very good areas of ling heather at Ord of Caithness, Lingwell and Braemore, some patches of *Erica cinerea* and cross-leaved heath in all areas.
52. Mountains of Mourne, County Down. Good areas of *Erica cinerea* with minor areas of ling heather.
53. Sperrin Mountains, County Tyrone and Derry. The ling heather areas are patchy. Some areas of *Erica cinerea* growing with cross-leaved heath.
54. Omagh, County Tyrone. Sixmilecross. Areas of ling heather.
55. Cooley Mountains, County Louth. Good areas of bell heather with minor areas of ling heather.
56. The Bog of Allen, central Ireland. Produces much ling heather.
57. Dublin Hills, County Dublin. Ling heather and areas of *Erica cinerea* growing with cross-leaved heath.
58. Wicklow Mountains, County Wicklow. Good ling heather. Areas of *Erica cinerea* growing with cross-leaved heath.
59. Connemara, County Galway. Good ling heather. Areas of *Erica cinerea* growing with cross-leaved heath.
60. Mountains of Kerry, County Kerry. Very good ling heather. Areas of *Erica cinerea* growing with cross-leaved heath.

The Heather Moorlands of the British Isles

There are numerous heather areas on blanket bogs in other parts of Northern Ireland that are largely ling.

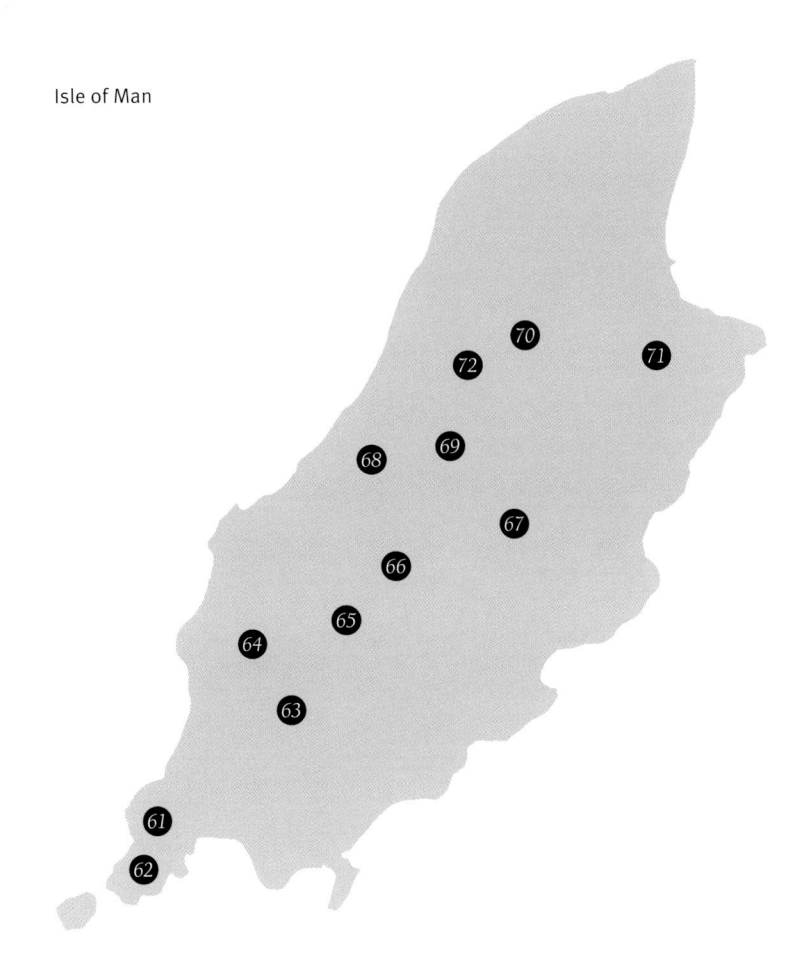

Isle of Man

91

61. Bradda. Patchy ling heather. Reasonable areas of *Erica cinerea* growing with cross-leaved heath.
62. Cregneash. Patchy ling heather. Reasonable areas of *Erica cinerea* growing with cross-leaved heath.
63. South Barrule. Ling heather. Reasonable areas of *Erica cinerea* growing with cross-leaved heath.
64. Dalby Mountain. Ling heather. Reasonable areas of *Erica cinerea* growing with cross-leaved heath.
65. Lower Foxdale. Ling heather. Reasonable areas of *Erica cinerea*.
66. Greeba Mountain. Ling heather. Reasonable areas of *Erica cinerea* growing with cross-leaved heath.
67. Injebreck Colden. Ling heather. Reasonable areas of *Erica cinerea* growing with cross-leaved heath.
68. Cronk-y-Voddy and Lambfell Moar. Ling heather. Reasonable areas of *Erica cinerea* growing with cross-leaved heath.
69. Sartfell and Slieau Freoaghane. Ling heather. Reasonable areas of *Erica cinerea* growing with cross-leaved heath.
70. Slieau Managh and Sulby. Ling heather. Reasonable areas of *Erica cinerea* growing with cross-leaved heath.
71. North Barrule. Ling heather. Reasonable areas of *Erica cinerea* growing with cross-leaved heath.
72. Slieau Curn and Glen Dhoo. Ling heather. Reasonable areas of *Erica cinerea*.

FUTURE CONCERNS FOR THE MOORLANDS

Areas of heather moorland throughout Great Britain are suffering an invasion of softwood forestry. The various agencies with interests in forestry have vast financial, technical and labour resources and are creating new plantations that, over time, will quickly see the list of areas above become dated.

It is said that oak and pine make up 59 per cent of British forests. In a report (August 2014), scientists at the University of Twente and the Forestry Commision's research agency highlighted that climate change may see native species of oak and pine in our woodlands replaced with trees from the continent such as Japanese red cedar and giant redwoods. Species grown in Victorian times are also expected to return as foreseers endeavour to ensure Britain's woods can survive rising temperatures and increased frequency of drought.

The Forestry Commission has indicated that the results of the study underpin plans for foresters to safeguard Britain's forestry areas. A spokesman, John Weir, advised that current species of oak are not doing very well and that we need to look at the range of alternatives available. Existing tree species from southern Britain will be migrated northwards, with plans to replace them with species from the Continent. For example, he stated that the North York Moors are likely to have drier summers which will not suit spruces and that, in the south east, we might grow some of the south-European tree varieties. Giant redwoods, which are already planted across the UK, are most likely to be used on the east coast, which will become warmer. The concensus of experts is that if nothing is done, it will be bad news for our natural woodlands.

THE HEATHER HONEYS

There are few data available regarding the constituents of heather honey other than that its protein colloids can be in excess of 1.85 per cent compared with those of most honeys at 0.2 per cent. In addition, ling heather honey is known to have a water content that can be as high as 24 per cent. This high water content is due mainly to the bee's inability to reduce the amount of water in the nectar by evaporation because of the nectar's natural thixotropic properties. Much research on honey was undertaken in the late 1940s by the late J Pryce-Jones whose study, *Rheology of Honey*, was published in 1953. Since that time, further research regarding honey, including heather honey, has been patchy.

Pressed thixotropic ling heather honey in a pair of jars showing the characteristic distribution of air bubbles [Brian Nellist]

Ling heather honey is unique amongst all honeys in colour, taste, viscosity and thixotropy. Thixotropy is the property exhibited by certain gel-like substances of becoming liquid when stirred; these substances will gel again if left to stand and the process can be repeated. This characteristic property of ling heather honey is thought to be because it contains more colloids than other honeys.

Ling heather honey has a high mineral content that has an important influence not only on flavour, but also on such factors as colour and the rate of wild yeast growth.

The acids in honey have a great influence on its taste. Malic acid and citric acid are the principal ones present, along with traces of acetic, succinic and formic acids. Besides imparting a tart taste, these acids affect the flavouring substances present.

Substances responsible for a honey's acidity amount to 0.1 per cent. When tested by certain methods, ling heather honey, with its high mineral content, does not show as high an acidity as other honeys of lower mineral content. The mineral constituents act as a so-called buffer.

In the case of ling heather honey, the acid and mineral content has an important bearing on the taste which is commonly agreed to be distinctive.

CONTENTS OF A FLORAL HONEY

Honey is a mixture of sugars and other compounds that make it unique as a food. It is easy for humans to digest because when the honey bee takes up nectar, enzymes and complex organic substances are added which invert the sugars. Other elements are introduced and the nectar is processed into honey which contains valuable, easily assimilated nutrients.

The nutritionist is most interested in the 3–4 per cent of 'other substances' within the honey. These can vary in quantity from one honey to another which, to the nutritionist, makes honey somewhat unique.

The constituents of a typical sample of honey are:

- water 18–21 per cent
- glucose (dextrose) 32–35 per cent
- fructose (levulose) 38–40 per cent
- other sugars (sucrose, maltose) 4–5 per cent
- other substances (vitamins, minerals, antioxidants) 3–4 per cent.

The other substances can include:

- up to 16 organic acids including gluconic, acetic, malic, succinic and butyric acids
- up to 12 mineral elements including potassium, calcium, sulphur, iron and chlorine
- up to 17 free amino acids including proline, lysine and glutamic acid
- vitamin C
- about 4–7 per cent proteins.

Other characteristics relating to honey which are not generally understood are density and viscosity. These two terms, which mean entirely different things, are not necessarily connected and are frequently used incorrectly.

Density means the relative weight. For our purposes it can be considered that density and specific gravity (SG) have practically the same meaning. Therefore, for example, when

The Heather Honeys

The nutritional value of honey per 100 g (3.5 oz)

Energy	1,272 kj – 304 kcal	Vitamin B2 – riboflavin	0.038 mg
Carbohydrates	82.4 g	Vitamin B3 – niacin	0.121 mg
Sugars	82.12 g	Vitamin B5 – pantothenic acid	0.068 mg
Dietary fibre	0.2 g	Vitamin B6 – pyridoxine	0.024 mg
Fat	0.8 g	Vitamin B9 – folate acid	2 µg
Protein	0.3 g	Vitamin C – ascorbic acid	0.5 mg
Water	17.10 g	Iron	0.42 mg
Calcium	6 mg	Phosphorus	4 mg
Magnesium	2 mg	Potassium	52 mg
Sodium	4 mg	Zinc	0.22 mg

Source: USDA Nutrient Database

analysis shows that a particular honey has an SG of 1.250 it indicates that, at normal temperatures, when ten litres of water would weigh 10 kg, the same volume of honey would weigh 12.5 kg. The honey is more dense; that is, heavier than its equivalent volume of water.

Viscosity is a measure of the resistance to flow which a liquid exhibits and changes with temperature. In practice, the higher the temperature (within limits), the less the resistance and the more 'fluid' the liquid. When a jar of honey is examined, it may be turned upside down and the honey's behaviour noted with regards to flow. The comment in this case is not about density (relative weight) but about viscosity, eg, 'a good, viscous honey'.

That notwithstanding, laboratory testing, as opposed to field work, employs delicate precision instruments to determine both density and viscosity to a high degree of accuracy. The beekeeper or honey judge cannot gauge density or viscosity accurately, but can turn honey jars upside down to provide an excellent guide to the latter.

The specific composition of any batch of honey depends on the flowers available to the bees that produced the honey.

BELL HEATHER HONEY

Bell heather is rarely found in large expanses on its own, although there are areas that still exist on isolated moorland. Generally, most bell heather is established in small areas, in tufts, or intermingled with ling heather. However, there are areas in Dorset that have large expanses which can be harvested.

The colour of bell heather honey varies from port wine to dark brown; its colour is influenced by the nature and acidity of the soil where the plants grow. Bell heather honey generally darkens and crystallises as it ages. Crystallisation is a relatively slow process and produces unmistakable large coarse flint-type crystals. The aroma of bell heather honey is pronounced. The flavour is strong, very often with a woody almond taste. The soil where the plant grows often imparts a slight minty, bittersweet taste to the honey that varies from location to location. Cross-blending of flavours arises from the intermingling of nectars obtained from bell heather and the cross-leaved heath. Bell heather is readily extracted from the comb by centrifugal extraction methods, whereas ling heather requires special treatment.

Bell heather honey from the heathlands of Dorset. Note the reddish orange hue that can be more a port wine colour from other moorland areas [Michael Badger]

LING HEATHER HONEY

Of all honeys native to the British Isles, ling heather is the most unique. It is the most viscous of all honeys and cannot generally be extracted in the usual way by centrifugal force, unless the honey is agitated in the comb cells beforehand. This process usually requires a special device with spring-loaded rods that are pushed into the capped cells of the honeycomb. The honey can then be extracted with a tangential extractor. Ling heather honey is ideally marketed as comb honey – cut comb, sections or Ross Rounds – avoiding the extraction process altogether.

Ling heather honey is a brilliant golden amber colour that can vary from reddish orange to dark amber. Again, the nature of the soil in which the plants grow influences the honey's colour and acidity. It imparts a slightly bitter, tangy, pungent,

Air bubbles in this pressed ling heather honey can be seen clearly [Brian Nellist]

smoky, mildly sweet taste that persists for a long time on the palate. The honey has a strong, distinctive, woody, warm, floral, fresh, fruity aroma, reminiscent of heather flowers. The crystallisation rate is quite slow in pure samples and in the early stages after extraction. Indeed, the honey may not crytallise at all. For less than pure samples, crystallisation results in a smooth light-coloured mass. The purest ling heather honey (a rare commodity) comes from higher-altitude moors where contamination from other floral honeys is likely to be minimal. Like all honeys, ling heather honey darkens in colour through natural ageing, or if stored in a refrigerator or freezer.

Ling heather honey, being thixotropic, is normally gel-like and firm but will become liquid temporarily if stirred or agitated. A common test of purity by honey judges is to place an opened jar on its side to see how quickly the honey will flow. A pure sample of heather honey will stay firmly in place for some time. Generally, the longer it stays in place the purer the sample. In competitions, the judge will strike a line across the surface of the honey with a glass rod to test its quality. How well the line retains its form is an indicator of how thixotropic the honey is and if it has been heated. The expert honey judge is quick to home in on this particular trait.

Some 37 years ago, I was judging heather honey at the Cleveland Show, North Yorkshire. Removing the lid from one particular honey jar, I immediately noticed the thumb print on the honey surface made by a previous judge. The honey's purity was further confirmed when the jar was laid on its side without any sign of movement of the contents.

A pressed sample as opposed to an extracted sample shows large air bubbles trapped in the gel-like honey, while honey extracted using a high-speed centrifuge introduces small uniform air bubbles throughout.

Another unusual characteristic of heather honey is its normally high water content. This can be as high as 24 per cent, but it is usually between 19 per cent and 23 per cent. The high water content encourages fermentation caused by wild yeasts naturally present within the honey. To kill off these unwanted yeasts, a method of pasteurisation is recommended.

CHOOSING A MOORLAND HEATHER SITE

Beekeepers new to heather honey production very often underestimate the importance of choosing a good heather site. Such a site is one of the three 'musts' for a successful harvest. The others are strength of the foraging force and good weather. It goes without saying that accessibility must not be overlooked. William Hamilton was known to state in his lectures and in his books that the best heather grows in Scotland and with that, the best heather honey comes from there, too. It is an argument that bears no real scientific basis, especially when there are large tracts of well-managed heather moors throughout the British Isles. It is true that Scottish heather honey has an excellent fine flavour, although some food connoisseurs would argue that other localities produce fine heather honeys of equal merit.

Experienced heathermen are adept at determining the likely quantity of the ling heather crop. William Hamilton states emphatically: 'The purple of the heather has more red in it when it is good'. Old hands can tell before the flowers open what the prospects of a good surplus are likely to be. The farming axiom that 'a wild and blustery May fills the barns with corn and hay' applies equally to heather. Dry periods and drought in May affect the heather to the extent that the plant foliage may never recover from the effects of water shortage. A dry period also gives the heather beetle a free hand. Over the coming years, the ever-worrying concerns regarding climate change will, perhaps, create issues relating to the particular conditions needed to sustain ling heather.

ACCESSIBILITY OF THE HEATHER MOORS

Above all, an apiary site should be accessible for moving hives onto a moorland stance and off again. Reliance on four-wheel-drive vehicles should not override common sense when working in rough terrain and where weather changes quickly. Difficulties may be encountered including vehicle breakdown, getting bogged down in soft ground conditions, or even injury. All these considerations must not be overlooked. A mobile phone signal may not be available to summon help when it is most needed. It is suggested that beekeepers should never work alone on moorland sites.

POSSIBLE SITES

It is essential to reconnoitre possible heather sites during early summer. A visit to the moors to ask permission from the moorkeeper or landowner should always be undertaken well in advance. Permission is rarely, if ever, refused. Do not put hives down on the moor, no matter how isolated the site, without seeking permission first. Doing so creates potential difficulties for all beekeepers that have built trust with the moorland stakeholders. Much information can be gleaned from sheep farmers and gamekeepers. The beekeeper should always be respectful to the landowner when taking his or her hives to a moorland site by not churning up the ground with over-zealous use of a four-wheel-drive vehicle.

A good spot for a small number of hives can be a forester's or shepherd's garden, naturally close to the moorland heather. A few jars of honey is the rent needed for such a site. Wherever a site is chosen, sheep are likely to be grazing, so this needs

A commercial beekeeper's beehives ready for unloading [John Mellis]

Polystyrene hives placed on a heather stance with hive straps in place to ensure stability in strong winds and to protect them from displacement by livestock [John Mellis]

to be borne in mind when the hives are set down. Sheep will invariably use hives as scratching posts which can be calamitous for the beekeeper if they are not securely strapped while they are on the heather stance.

SELECTING A SUITABLE HEATHER SITE

The amount of heather growing in the area should also be considered. The chosen site should have at least 40 plus hectares (100 acres) within immediate foraging range of the bees in dense, mass areas. Small patches of heather are of little use.

Ling heather honey is seldom, if ever, produced in quantity from plants growing on deep peaty soils that are constantly wet, or land that is continually waterlogged. There are exceptions at times of drought when the nectar will flow.

Look for those sites which are on actively managed, driven grouse moors. These moors are often burnt periodically under controlled conditions; this ensures the old straggly heather is burnt off, soon to be replaced with young heather that quickly flourishes. Such moorland burning is undertaken by expert moorland managers whose knowledge has been handed down over the generations. In addtion to feeding the grouse, such heather plants have the added benefit for the bees and the beekeeper in that they are in the ideal condition for nectar secretion.

The burning off of old heather plants is also a method for controlling the dreaded heather beetle, *Lochmaea suturalis*, native to north-west Europe. It feeds upon heather foliage. Heather beetles are difficult to spot as they are camouflaged by their brownish colour. The beetle is about 6 mm (0.24 in)

A sheltered site with water nearby and abundant heather [John Mellis]

long. They have a tendency to hide and they drop into the undergrowth if they are disturbed. They are easier to see when present in large numbers on the same plant. The heather beetle eventually kills the plant. Browning of the plant's foliage that mimics frost damage is caused by the beetle's feeding.

SITE CONSIDERATIONS FOR A HEATHER STANCE

The ideal location is between sea level and 300 m (985 ft). A shallow depression that shelters hives from westerly winds is essential; storm watercourses or natural drains from the surrounding hillside should be avoided. Well-sheltered hollows are useful provided that they are not in deep shade throughout the day as such sites harbour cold air. A site that has dark chilling shadows falling across the hives, no matter what time of day, will shorten the working day of the foraging

A site where rainwater may collect is to be avoided; note the foreground vegetation which indicates rainwater may collect here [Anne Lever]

An ideal moorland situation that gets the sun from dawn to dusk. It is sheltered from the westerly wind, is a free-draining site and has easy access from the road, making positioning and moving hives more practical [Anne Lever]

bees. A poor harvest is more often caused by poor bee-flying weather than by failure of nectar secretion. The beekeeper must maximise the length of time available each day for bees to forage.

Choose a site where the bees do not have to labour to get the crop back to the hive. The bees should not have to fly uphill but should return to the hive fully laden downhill, coasting on the wing. Avoid setting a site up near a busy road, where the bees need to traverse it to access the heather. Thoughtfulness to walkers and horse riders should not go amiss; ensure beehives are kept well away from rights of way or green lanes. Vandalism may occur if hives are easily accessible.

Importance of Altitude

Altitude plays a major part in coastal regions, with intermittent fog and sea frets occurring in areas such as the north-west and north-east of England and Scotland. Heavy rains can and will wash the nectar from heather flowers, curtailing the flowering period. In my 56 years of heather-going experience, my records from 1959 to 2014 show the average good weather spell during which the bees can obtain pure ling nectar is, at best, 10–12 days. Towards the end of August, the bees will forage for other crops (eg, late clover) which will result in a blended honey. Ling heather is renowned for its low nectar-yielding potential compared with other major honey-producing flora.

In my talks with the late Colin Weightman and the late Brother Adam, there was general agreement that the further the bees have to fly to the heather, the less able the foragers

Snow on a Tayside hillside, Braemar, Aberdeenshire, mid May 2013. Well-sheltered hollows are notorious cold spots; away from the snow gully the temperature was 16 °C [Erica Osborn]

are to amass honey quickly. This distance factor can be most important with regards to filling heather supers in changeable weather. Common sense says the placing of hives nearer to the forage enables the bees to fly shorter distances, thus allowing more flights per day and, consequently, more nectar gathering with less wear and tear on the bees. So, the bees should be located in the heart of the moor or on mountainsides that, above all, are abundant with heather with a spread that is not at all inconsistent or patchy.

Mountainous Areas

Above altitudes of 300 m (985 ft) above sea level, the ambient air conditions rarely help the beekeeper. Early frosts are commonplace from mid August. In mountainous locations, the chosen site should permit the bees to fly from the hive to higher ground, returning to the hive by coasting on the wing. Many beekeepers find a sheltered valley with heather on both sides as a means of ensuring maximum foraging area. Hive entrances are best directed to the east to catch the early morning sun, also avoiding cold north-west winds; it is not desirable to have a cold wind blowing into the hive entrance. The maximum altitude for hives should be 650 m (2130 ft) above sea level; the lower the altitude the better.

Scrub Farmland and Softwood Forestry Plantations

Avoid the temptation to find a site near scrub farmland. Such areas are often accessible and near to the roadside. They consist of pasture land that all too often harbours the notorious weed ragwort (*Senecio jacobaea*), a yellow daisy-like flower with a flowering period from mid July to the end of summer. Its nectar produces a bitter honey and it takes only a very small amount to spoil what might have been a splendid sample of heather honey.

Common ragwort (also known as benweed, staggerweed, tansy ragwort and St James's wort) is probably the most common of the poisonous plants growing on roadsides, waste ground, pasture and agricultural land in Britain. It is one of the five species listed as a noxious weed in the Weeds Act 1959, which requires the landowner to prevent ragwort from spreading on his or her land, or land under his or her control, when served notice to do so.

Forestry Commission land can be a haven for rosebay willowherb (*Chamerion angustifolium*) in forestry clearings,

Scrub farmland. Ragwort (*Senecio jacobaea*), a yellow daisy-like flower, flowers in the late summer. West End, North Yorkshire [Michael Badger]

A large expanse of rosebay willowherb (*Chamerion angustifolium*) near to a forestry clearing at Fylingdales, North Yorkshire, late July 1961 [Author's collection]

especially in rides where there have been fires. Rosebay willowherb thrives on potash-rich soil. This plant is known in the Americas as 'fireweed' because of its liking for such ground conditions. Its nectar contaminates the ling heather, leading to flint-like granulation that is easily detected in a sample after a few weeks. Such honey should not be discarded but marketed as a heather blend. Other plants to be avoided include wood sage (*Teucrium scorodonia*) and blackberry (*Rubus fruticosus*). Bees will forage on these plants, turning a good sample of heather honey into a blended moorland honey that granulates quickly with clearly visible flint-like granules. Although eaten in quantity, the connoisseur will always prefer the purer sample over its blended counterpart.

Honeydew

There are times and instances when the bees will collect honeydew from woodland areas that abut the heather moorland. This foraging activity may arise as a result of poor-yielding heather that instigates foraging elsewhere. Honeydew is a sticky, sugary, dew-like exudation found on the leaves and stems of trees and plants, especially in dry warm conditions. There are two kinds of honeydew: one is directly obtained from plants and trees through broken plant tissue, the other is secreted by aphids. This secretion is of a saccharine substance produced in very large amounts that attracts honey bees in times of dearth. Honeydew from spruce (*Picea* spp.) or Douglas fir (*Pseudotsuga menziesii*) is a dark sweet liquid that bees will collect at will.

As is life, there is always an exception. Some 25 years or more ago, the late Herbert Pierson who lived on the heather

Unmanaged heath and heather growing alonside a young tree plantation at Cannock Chase, Staffordshire [Michael Badger]

moorland at Hartoft Dale, Rosedale, north-east Yorkshire, had a quantity of an admixture of honeydew, bell heather and ling heather that was extremely palatable. The honeydew came from nearby spruce trees and had a very dark port wine colour that would have sent honey judges into a huddle trying to determine its origin.

On the whole, honeydew spoils a good honey, so forewarned is forearmed.

Use of the Same Heather Stance
Year after Year

Over the years, I have found that the same site is not necessarily the best year after year because of periodic burning, whether by keepers of well-managed moors or through accident or vandalism. The amount of suitable

Hives on a burnt moor with smoke from fires in the far distance, Saltersgate Moor, North Yorkshire [William Slinger]

heather available may vary enormously as a result of these actions. As mentioned previously, a pre-visit to the moors in early July gives time to select another site should it become necessary. It is good practice to have a spare site in reserve for such eventualities. Effective liaison with moorkeepers pays dividends as they will help you to find good sites to obtain a worthwhile crop.

Soil Conditions Where the Heather Grows for Maximum Return

The underlying bedrock of upland moorland influences the natural soils, vegetation, animal presence and abundance of grouse.

According to the late William Hamilton, Joseph Tinsley and Alexander SC Deans, the best conditions for ling heather

Choosing a Moorland Heather Site

are on shallow acid soils derived from igneous metamorphic rocks of whinstone and greenstone. The heather plant seems to yield the finest and most abundant nectar on moderately thin soil over bare rock. It needs little moisture to yield good quantities of nectar under such scant conditions so that the bees can work it. Warm sunshine and little wind during the middle of the day is all that is required in addition. The avoidance of wet bog land is essential for the maximum yield of nectar.

Aspect

Aspect should be carefully considered in terms of potential damage to the heather blossom by late frosts in May and

High altitude heather at 610 m (2001 ft) in mid afternoon with cloud, mist and no sun. Royal Deeside, Ballater, Cambus O'May, Aberdeenshire, 1962
[Michael Badger]

early June, a crucial time just when the buds are forming. The heather plant is very hardy; it will stand severe cold and biting north-easterly winds. However, its flower buds are tender and easily injured by such conditions. If this happens, a poorer yield can be expected unless nature takes a hand with ideal growing conditions at a later date. As mentioned previously, heather is adapted to yield nectar at relatively low temperatures in its moorland habitat.

Frosty conditions at Beeley Moor, Chatsworth, Derbyshire

[Trevor Marsden www.tezmarsdenphotos.co.uk]

SECTION 1
SUGGESTED FURTHER READING

- Allen, MY (1936). *European Bee Plants and their Pollen*. Alexandria, Egypt, Bee Kingdom League.
- Ashley, M (1964). *Life in Stuart England*. London, BT Batsford Ltd.
- Atkins, W (2014). *The Moor: Lives, Landscapes, Literature*. London, Faber and Faber Ltd.
- Avery, M (2015). *Inglorious: Conflict in the Uplands*. London, Bloomsbury Publishing Plc.
- Bagley, JJ (1960). *Life in Medieval England*. London, BT Batsford Ltd.
- Berry, RJ (1977). *Inheritance and Natural History – No 61*. London, New Naturalist Series, Collins.
- Birley, A (1964). *Life in Roman Britain*. London, BT Batsford Ltd.
- Butler, CG (1954). *The World of the Honeybee – No 29*. London, New Naturalist Series, Collins.
- Cecil, R (1969). *Life in Edwardian England*. London, BT Batsford Ltd.
- Chamberlin, ER (1972). *Life in Wartime Britain*. London, BT Batsford Ltd.
- Cowan, TW (1908). *Wax Craft*. London, Sampson Low, Marston & Co Ltd.
- Crane, E (1975). *Honey: A Comprehensive Survey*. London, William Heinemann Ltd.
- Crane, E (1983). *The Archaeology of Beekeeping*. London, Gerald Duckworth & Co Ltd.
- Cullen, LM (1968). *Life in Ireland*. London, BT Batsford Ltd.
- Darling, FF (1947). *Natural History in the Highlands and Islands – No 6*. London, New Naturalist Series, Collins.
- Digges, Revd JG (1904). *The Practical Bee Guide*. (Originally titled *The Irish Bee Guide*). Dublin, The Talbot Press Ltd.

- Dodd, AH (1961). *Life in Elizabethan England*. London, BT Batsford Ltd.
- Dummelow, J (1973). *The Wax Chandlers of London*. London, Phillimore.
- Dyer, C (2002). *Making a living in the Middle Ages: the people of Britain 890–1520*. New Haven and London, Yale University Press.
- Fitter, RSR (1945). *London's Natural History – No 3*. London, New Naturalist Series, Collins.
- Fleure, HJ (1951). *A Natural History of Man in Britain – No 18*. London, New Naturalist Series, Collins.
- Fraser, HM (1931). *Beekeeping in Antiquity*. London, University of London Press Ltd.
- Free, JB and Butler, CG (1959). *Bumblebees – No 40*. London, New Naturalist Series, Collins.
- Gardiner, J (2005). *Wartime: Britain 1939–1945*. London, Review.
- Gilmour, J and Walters, M (1954). *Wild Flowers – No 5*. London, New Naturalist Series, Collins.
- Gregg, AL (1949). *The Philosophy and Practice of Beekeeping*. Petts Wood, Bee Craft Ltd.
- Hamilton, W (1945). *The Art of Beekeeping*. 3rd Edition 1946. York, The Herald Printing Works.
- Hart-Davis, D (2015). *Our Land at War. A Portrait of Rural Britain 1939–45*. London, William Collins.
- Herrod-Hempsall, W (1930). *Beekeeping New and Old described with Pen and Camera – Volume 1*. London, The British Bee Journal.
- Herrod-Hempsall, W (1937). *Beekeeping New and Old described with Pen and Camera – Volume 2*. London, The British Bee Journal.
- Hodges, D (1952). *The Pollen Loads of the Honeybee*. London, Bee Research Association.
- Hoskins, WG and Dudley Stamp, L (1963). *The Common Lands of England & Wales – No 45*. London, New Naturalist Series, Collins.

Section 1 Suggested Further Reading

- Howes, FN (1945). *Plants and Beekeeping*. London, Faber and Faber Ltd.
- Imms, AD (1947). *Insect Natural History – No 8*. London, New Naturalist Series, Collins.
- Kesseler, R and Harley, M (2004). *Pollen*. London, Papadakis Publisher.
- Manley, ROB (1936). *Honey Production in the British Isles*. London, Faber and Faber Ltd.
- Manley, ROB (1948). *Bee-Keeping in Britain*. London, Faber and Faber Ltd.
- Page, RI (1970). *Life in Anglo-Saxon England*. London, BT Batsford Ltd.
- Patten, M (2005). *Feeding the Nation*. London, Hamlyn Books.
- Pearsall, WH (1950). *Mountains and Moorlands – No 11*. London, New Naturalist Series, Collins.
- Pettigrew, A (1870). *The Handy Book of Bees*. Edinburgh, William Blackwood and Sons.
- Priestley, JB (1942). *British Women Go to War*. London, Collins.
- Proctor, M and Yeo, P (1973). *The Pollination of Flowers – No 54*. London, New Naturalist Series, Collins.
- Pryor, F (1998). *Farmers in Prehistoric Britain*. Stroud, Tempus.
- Pryor, F (2010). *The Making of the British Landscape*. London, Allen Lane.
- Pryor, F (2011). *The Birth of Modern Britain*. London, Harper Press.
- Rackham, O (1986). *The History of the Countryside*. London, JM Dent and Sons Ltd.
- Reader, WJ (1964). *Life in Victorian England*. London, BT Batsford Ltd.
- Rebanks, J (2015). *The Shepherd's Life: A tale of the Lake District*. London, Allen Lane.

- Seaman, LCB (1970). *Life in Britain between the Wars*. London, BT Batsford Ltd.
- Seeley, TD (1995). *The Wisdom of the Hive*. London, Cambridge, Massachusetts, Harvard University Press.
- Seeley, TD (1985). *Honeybee Ecology*. Princeton, Princeton University Press.
- Seeley, TD (2010). *Honeybee Democracy*. Princeton, Princeton University Press.
- Smith, C (1992). *Late Stone Age Hunters of the British Isles*. London, Routledge.
- Stamp, LD (1976). *Britain's Structure and Scenery – No 4*. London, New Naturalist Series, Collins.
- Tautz, J (2008). *The Buzz About Bees*. Berlin, Springer-Verlag.
- Tomkeieff, OG (1966). *Life in Norman England*. London, BT Batsford Ltd.
- Weightman, C (1961). *The Border Bees*. Consett, Ramsden Williams Publications.
- Weightman, C (1991). *Tales of a Border Beekeeper*. Shilford, C Weightman.
- White, RJ (1963). *Life in Regency England*. London, BT Batsford Ltd.
- Winston, ML (1987). *The Biology of the Honey Bee*. London, Cambridge, Massachusetts, Harvard University Press.

SECTION 2:

HEATHER HONEY PRODUCTION

'Going to the heather moors is a well-established form of migratory beekeeping for many beekeepers. With old hands, preparation and fastening up of colonies, their transportation to the moors and the release of the bees when the hives have been located is a clearly defined ritual, never varying from year to year in the smallest detail, and the first light of the first Sunday in August sees the outward trek of beekeeping nomads in co-operative parties, tired but merry as they journey out, more tired but even merrier on return.'

[Recorded by the Revd K Harper, editor of *Northern Apiarist*, from the lecture given by Alfred Hebden, NDB, County Beekeeping Instructor of the West Riding of Yorkshire, 6 May 1950, at a Cumberland and Westmorland Beekeepers' Association meeting at Gosforth, Seascale]

The late Colin Weightman, MBE, in August 2009 removing the heather crop. Shilford, Northumberland [David Pearce]

STRAINS OF BEES FOR HONEY PRODUCTION

Over the past 40 years, there has been much debate amongst specialist groups about the 'best' strain of bee for the use of beekeepers. It could be argued that much of what has been said, published and lectured upon has been hot air. Other than that, many studies have taken place on wing venation and the practicalities of producing hygienic bees. Very little progress would appear to have been made to introduce or develop a strain of bees best to meet the needs of the beekeepers of the British Isles. However, the subject does draw beekeepers together with a common interest.

The disappointment, if there is one, is that there is no real detailed specification published that readily identifies the characteristics of the bee strains bred and developed by proponents of breeding better bees.

What is available to the beekeeping fraternity is progeny developed from the bees of other individual beekeepers or commercial ventures.

Under current European Union (EU) trading agreements, the General Agreement on Tariffs and Trade (GATT) and Bali world trading agreements, it is virtually impossible to ban the importation of bees into the United Kingdom and the Republic of Ireland. Yet as long ago as June 1944, the late Arthur Abbott, of Mountain Grey Apiaries, and nearby Yorkshire Apiaries at Willerby, Hull, supported such a moratorium on imports. Throughout 1945 and 1946, these businesses persevered with setting up an organisation to control such imports and to set up a national agency for the production of home-produced queens under a banner of 'The proposed British Bee-Breeders' Association'. I was given a paper by Arthur Abbott that he had produced with one of his fellow directors, Freddy Wilkinson, which, after much deliberation, was published nationally. It was made available to all the beekeeping journals and circulated to all their fellow commercial breeders. The paper (reproduced from *Bee Craft*, June 1944) outlined their aims and aspirations as follows:

> *Feeling the need for placing the bee-breeding industry on a more solid foundation, with more control in the supply of British reared bees and queens, we are of the opinion, that this object would best be served by the formation of an association, comprising all British commercial queen and bee-breeders, who would pledge themselves to work for that end.*
>
> *If the future is to be safeguarded, (1) from the viewpoint of the ultimate improvement of the bee by up-to-date methods of selective breeding, (2) to meet the threat of a possible large-scale 'flooding' of the country by foreign bees at cheap prices but of dubious value, with ever present risk of importing disease, and (3) in order to provide all reliable breeders with a definite status, and to keep faith with the small bee-keeper by offering him a guaranteed product, such an association would appear to be essential ...*
>
> *... Briefly the aims of the association are:–*
>
> a) *To work for the development and improvement of the honey-bee.*
> b) *To establish – through its members – stabilised strains of bees of proved worth, by selective*

Strains of Bees for Honey Production

> *breeding methods in accordance with modern knowledge of genetics, and that one of the essential qualifications of membership shall be that such a strain has been, or is in the process of being established by him.*
>
> c) *To register such member's strains as in the form of a 'pedigree' on sufficient evidence that such a strain is as represented by the breeder, who will give his personal guarantee that all bees and queens sold, or offered for sale by him, will be of this strain unless otherwise stated. Freedom from disease in the breeder's apiaries will also be an implied condition of sale. In this way the bee-keeper will be safeguarded, and the confidence in the products of British breeders will be enhanced.*
>
> d) *To seek to maintain the existing embargo on importation of bees and queens for as long a period as possible after the cessation of hostilities; or failing this, to secure limitation of imports for as long as possible, in which case powers would be sought to make it an offence to sell, or advertise for sale such foreign bees, without disclosure of their origin.*
>
> e) *To fix ruling prices for bees and queens to which all members will adhere, such prices to be subject to revision as and when deemed advisable.*
>
> f) *To meet at convenient intervals for the purpose of meetings and lectures, etc.*

The narrative ended with a plea for support.

AW Gale, Marlborough, wrote a response in *The Scottish Beekeeper*, July 1944:

> Sir, – As probably the largest breeder of bees in this country (we despatched nearly 1100 colonies during 1943) I do not agree with Messrs Abbott & Wilkinson in many of the proposals they put forward in their circular letter.
>
> Examining their aims in detail, the first is a laudible endeavour which every sound beekeeper would support. The second, however, though admirable in sentiment, appears to me to be most difficult to apply in actual practice. I have a large herd of pedigree Wessex pigs and also some hundred odd Ayrshire cattle, some pedigree and some non-pedigree, which latter I am trying to grade up by the use of high-class pedigree bulls. With cattle and pigs each animal is ear-marked at birth, breeding is controlled, and errors and frauds can be eliminated; but I submit that it is virtually impossible to suggest any practicable scheme which can guarantee fully controlled mating, without which 'selective breeding methods in accordance with modern knowledge of genetics' is impossible. The authors state that an essential qualification of membership shall be that such a strain has been, or is being, established; I maintain that it is impossible in the absence of identification marks and properly-authenticated records, to establish any such claim, and that any attempt to do so will merely open the door to fraudulent practices on the part of a number of 'get-rich-quick' beekeepers who will make use of the reputation of honest members to bolster up the sales of their own bees.
>
> The third item is subject to the same criticism; it is impossible to prove or disprove the breeding of any

particular strain of bee (as apart from a race) and without proof or capacity of proof a 'pedigree' means nothing. The only proof there would be the seller's own integrity; this is unaffected by the absence of a 'pedigree,' so why fool around with the latter?

I do not agree that it is either necessary or desirable to maintain the existing embargo on foreign bees; we are on controversial ground here, but my own strain of Caucasian hybrids originated from a queen imported from Alabama, and I have imported, and used for honey production, some thousands of queens, a large proportion of which were of very high quality (I regret to confess that in spite of every care, by no means all of my home-reared queens are of the highest quality; I try my best, but after 25 years I have not succeeded in rearing any number of queens all of which are equally good).

I also disagree with the fifth item; why should not each breeder fix his own price in open competition? If there is a flat price there will not be the same incentive to the breeder to produce a better article to gain a higher price. The effects of a controlled price are shown now-a-days in the marketing of fruit, etc.; as it is all the same price, new producers do not trouble about quality. – I am, etc.

Despite all their efforts, in the main, beekeepers from all fronts thwarted them as they wanted young queens to be available in April, early in the season, rather than later, in May. Advertisers used slogans like: 'Early swarms are the raw material for producing more British honey. They cannot be obtained in England in time to get the best results'. The beekeeping press at the time adopted a 'free press' approach to the subject, with the pros and cons being openly debated. Alas, feelings ran against such a move and the ideas for controlling queen and bee imports were inevitably kicked into the long grass of beekeeping history.

THE VILLAGE BEE-BREEDERS' ASSOCIATION

In 1964, Beowulf Cooper and Terry Theaker formed an organisation to establish a moratorium on importation of bees and queens by encouraging beekeepers to produce an apiary-bred strain. The Village Bee-Breeders' Association was to flourish and become the British Isles Bee Breeders' Association and now the Bee Improvement and Bee Breeders' Association (BIBBA).

A moratorium on imports would allow bee breeders to get to work on culling poor bee strains and stabilising the bee population towards a true natural strain that could be called *Apis mellifera mellifera* 'Britannica', or some appropriate title. In reality, the hard fact is that no serious-minded bee breeder would spend time, money and effort producing bees commercially that, in reality, would benefit competitors to his or her own disadvantage. Therefore, the reliance on voluntary bee breeding groups will continue, which is to be applauded.

The work of bee breeding groups (and there are many) is admirable. These enthusiasts are keen to ensure that new beekeepers are made aware that most strains of bees in these islands are no longer pure but are, in effect, mongrels. Any efforts to improve the strains of indigenous bees in the British Isles are to be encouraged. However, such efforts will continue to be thwarted until importation of bees is stopped.

BIBBA meeting at Mulgrave Castle, Lythe, North Yorkshire, 1976. (Back, left and right) Phil Lumley, Donald Foxton. (Front. left to right) John Dews, Jack Barker, Ben Taylor [Michael Badger]

Beowulf Cooper at the Great Yorkshire Show, 1981 [Michael Badger]

THOUGHTS ON BEE STOCK IMPROVEMENT

Having been involved in both farm animal and small livestock breeding from an early age, I quickly realised that stock improvement with cattle and other animals can be achieved in a relatively few years through a system of 'grading up', a process by which a sire (male) with the required characteristics is retained and a fresh dam (female), also with desirable traits, is brought in for breeding purposes.

The in-depth topic of animal genetics is beyond the scope of this book, other than to say that bee breeding is unique in nature and something special indeed. It is very difficult to control matings and, as far as I am aware, there have been no workable methods of mating a queen to selected drones in an enclosure. There is the possible forced method of instrumental insemination, but on what criteria is the drone being assessed when using this method?

With large animal breeding, the male and its characteristics are known to the breeder. The selection of a honey bee drone from desirable stock is a chance. It is not known whether the drone is a 'visitor' to the colony, as all drones are accepted into any hive during the normal season. This is probably the one true obstacle that confronts the honey bee breeder, to stay in control of both sexes for bee breeding.

The late Harold Woolhouse, professor of botany and zoological sciences at Leeds University and director of the John Innes Institute, Norwich, was both a beekeeper and a breeder of Brahma chickens. He stated to me quite emphatically that the honey bee, owing to its particular way of breeding, through parthenogenesis, has a somewhat unique position in terms of reproduction. The females (queens and workers) are developed from fertilised (diploid) eggs, whereas drones are developed from unfertilised (haploid) eggs. Drones have daughters but no sons and they have no father, only a maternal grandfather.

Owing to the manner of honey bee mating (in nature, on the wing), it is at best difficult, if not impossible, to control the process. It is known that queens will mate with possibly upward of a dozen drones in drone congregation areas. Experiments have been undertaken to control mating in an enclosed environment, but it seems that no real workable method has proved succesful.

Work has been undertaken in the Fresian Islands, the Netherlands, and in California, United States of America, whereby isolated mating stations, clear of undesirable drones, have been established. In the 1970s, I visited the Laidlaw Bee Research Centre, University of California, Davis, to see first hand work such as the Quinn–Laidlaw method of hand mating and Watson instrumental insemination. The basis of these experiments was that the sperm in an inseminated queen remain alive for several years and can be taken and transferred to another queen, thus making it possible to produce back matings by males (drones) that have been dead for some time. This procedure would enable the establishment and maintenance of the most desirable combinations of traits that are mainly derived by heredity.

Like all other animals in breeding scenarios, honey bees need to pass through a stage where progress is maintained by cross-breeding using a combination of purposefully selected genes. This, in the end, would result in bee breeders resorting to selection of stock according to genotype to fix the breed or strain. Once underway, the qualities of different bees can only be determined by comparison of a number of colonies.

To recap, each colony consists of the daughters of one mother so, in my opinion, selection falls back to genotype whereby testing for quality relies on progeny testing. This factor ought to make selection in cross-bred groups of honey bees very effective. My BIBBA colleagues would not necessarily agree with this statement. Views on the topic of genetics are, without doubt, very subjective; for my own part I have an open mind, other than to say that Professor Woolhouse was no slouch on such matters.

The Queen's Characteristics

A fact very often overlooked with honey bees is that the queen is not productive in herself (the queen is confined to the hive and does not forage but is, in essence, an egg-laying machine). The queen carries the quality of stamina. In addition, the only other real character that can be assessed with certainty is her fertility, as stated previously.

The Importance of the Drone

The drone's importance is paramount to the improvement of characteristics of honey bee colonies for honey production.

From the outset, a good supply of drones is needed. All mating nuclei should have their own drones as part of their make-up, even though drones will move freely between colonies. To this end, it is essential to rear drones with positive characteristics. It is of vital importance that any colonies of bees present in the apiary that are known to be bad tempered, or have other undesirable traits, should be dealt with. These stocks should be moved well away to another apiary, or the

Strains of Bees for Honey Production

queen replaced with one raised from a colony which exhibits all the desirable characteristics. Sub-standard colonies should not be permitted to have their drones 'on the wing'. Queen matings with drones from such colonies will propagate the undesirable qualities in the progeny. That will not do. The drone passes on the undesirable characteristics, in addition to any that are desirable, to both male and female castes.

The late Beowulf Cooper and Terry Theaker made great play on the importance of the drone. Their lectures were very often on the subject of the 'misaligned drone', emphasising the importance of breeding drones to the betterment of future stocks.

With the arrival of varroa, the drone has become even more maligned than it was previously. This is because of the drone's unfortunate interrelationship with this parasite. The mite is attracted to drone brood cells in preference to worker brood cells for the reproductive stage of its life cycle.

The late John 'Eddy' Eade, of Mountain Grey Apiaries in the East Riding of Yorkshire, flooded the surrounding area of Holme-on-Spalding-Moor with both queens and drones from his apiaries and supplied beekeepers in the neighbourhood as a means of keeping the Mountain Grey strain as pure as possible. This prolific strain of bee, developed by company founder Arthur Abbott, was bred successfully from the late 1920s and continued to be more or less pure well into the early 1980s. The success of these bees was due, in part, to devoted apiarist, Ken Jackson (octogenerian), and special adviser, the late Freddy Wilkinson. Isolated mating apiaries around North Cave, East Riding of Yorkshire, were also essential to this success.

Mention has already been made that, in the case of domestic farm stock, the male is mated to many females. Thus the male is of greater economic importance than the female. The honey bee drone is somewhat different in these terms:

- A drone will mate once only. This limits his influence.
- A drone is produced parthenogenetically; he has no father. The drone's hereditary qualities are those of his mother (queen) which, in turn, are those of her father and mother. Therefore, the drone inherits the characteristics of his grandfather and grandmother, which are collective in his mother.
- There are really no discernable direct means of assessing the transmissible qualities of a drone, apart from physical virtues such as colour.
- Instrumental insemination and isolation matings are the only methods of control of the characteristics of the offspring resulting from honey bee mating. Bees of different races kept in the same locality will have a marked predisposition to crossing. Instrumental insemination is not totally reliable owing to the fact that the chosen drones must be known to come from desired stock. I know that Brother Adam often marked newly emerged drones at his Sherberton apiary in the wilds of Dartmoor as a means of identifying their origins. When looking for a supply of drones it was interesting to note how itinerant they were, often appearing in a hive other than that from which they originated.

Strains of Bees for Honey Production

Drone rearing is easier than queen rearing, but is anything but motivating or fulfilling. It goes without saying that considerable effort is needed to make certain a supply of choice drones is available throughout the breeding season. The serious queen breeder does not leave drone production to nature.

In a visit to Harvey York Jr, Jesup, Georgia, USA, in 1979, I saw for myself how considerable numbers of drones were raised from the best stocks by placing several frames of drone comb into these colonies right through the season. At the same time, the amount of drone brood in the other stocks was restricted, knowing that each and every stock needs to produce its own drones to harmonise colony cohesion; colonies will not prosper without having produced their own drones. It is a fallacy to try to eliminate them.

From early April, stocks for rearing drones with all the desirable characteristics are chosen. A frame is inserted in each brood chamber with a 'starter' of foundation or, better still, a frame of drone comb. The starter is drawn as drone comb, especially if the stock is fed. A second frame with a starter or drone comb can be added, followed by a third. This process results in a large number of drones which will be present throughout the active season. The issue of swarming never arises because of the presence of large numbers of drones at the expense of workers. A colony must be balanced for swarming to be initiated. The reduced proportion of worker brood will ensure the colony never gets strong and will never produce surplus honey. As a result of its unbalanced state, this stock of bees puts a tremendous strain upon its nurse bees. Feeding may be essential to ensure that brood is properly nourished during its development – an essential requirement for drones to be vigorous and healthy for their paternal duties.

Cross-bred Groups of Honey Bees

Such methods ought to make selection in groups of cross-bred honey bees quite effective. The breeder needs to select those bees in his apiary that have desirable traits and propagate from them.

At one time, those stocks with undesirable traits would have been culled to ensure that consistency was maintained. Genetic theory and modern bee breeding no longer support this approach. For example, a high level of honey production in cross-bred stocks can be due to hybrid vigour, but propagation of these hybrids does not ensure that this trait will be continued in subsequent generations, as the genes become recombined in multitudinous ways with outcomes that are quite unpredictable.

To digress, with the breeding of pedigree cattle, we have found that, despite their purity, a rogue trait can easily surface. A beekeeper might easily be tempted to breed from stocks that are prolific nectar gatherers. However, this characteristic might well be the result of a robbing trait, rather than one of foraging. Really, one may never know.

Naturally, if you are working with a pure strain, this is the only one you are likely to keep consistent as there is a limited gene pool. With these bees, if you breed from the best, the offspring are typically as good as the parents, or even better. If, as a bee breeder, you conserve a restricted gene pool and breed from within it, then the outcome is predictable within the limits specified by that gene pool.

Strains of Bees for Honey Production

These limits can be extended by careful trial and error. If you mate these bees with others of a similar type, you are very unlikely to suffer a catastrophic outcome. By contrast, crossing with a different strain risks introducing maladaptive genes. In the first generation there may be spectacular results, because of hybrid vigour, but subsequently everything goes haywire and you end up with unproductive and, typically, highly aggressive bees that are also disease-prone, as they are not naturally selected for the particular environment. So forewarned is forearmed.

Temperament Groups

The late Beowulf Cooper and Terry Theaker drew attention to the observation that matings between the Mediterranean bees (ie, *A. m. ligustica*, *A. m. macedonica*, *A. m. carnica*, etc) need not result in bad temper. Similarly, matings between the different strains of *A. m. mellifera* also generally produce gentle offspring. However, crosses between the *A. m. mellifera* subspecies and any of the Mediterranean group can produce highly aggressive offspring. Cooper described this observation as indicating two different 'temperament groups', which you ignore at your peril.

    Over the years I have met many commercial beekeepers who, to say the least, are very modest in their successes. Such beekeepers include Keld Brandstrup, the leading Buckfast breeder in Denmark, who runs over 400 colonies for honey production, and Ged Marshall, a commercial beekeeper from Buckinghamshire, England. My own experiences come down to the old adage (and it really rings true for the average beekeeper) that 'the best bees are those bred in your own apiary'.

THE BEES IN ORKNEY

In August 2013, I was visited at my home apiary by Gavin Jones, an enthusiastic beekeeper from the Isle of Harris, Outer Hebrides. Gavin is one of several beekeepers on these islands. He was keen to see my bee house as a potential means to develop his own bee breeding project. The natural environment on Harris is windy, but the island has a reasonable climate with plenty of forage. The bees are free of varroa. There are no bee imports. The set-up that he has created makes me think that he is sitting pretty. The challenge is to breed bees true to type for stock improvement and for export. I wish him well with his endeavours.

The Author's bee house, Roundhay, Leeds, West Yorkshire [Natalia Fajczyk]

HEATHER BEES

The late Colin Weightman adapted his own strain of bees modelled on the north-European dark bee (*Apis mellifera mellifera*). Many of the dark bees in the United Kingdom stem from the extensive restocking scheme which followed the end of the Great War with imports of Dutch, Ligurians and French black bees. I attended a lecture given in the early 1960s by John Mills, former County Beekeeping Instructor, at a weekend school at Newton Rigg, Cumbria. He advised that many of the best comb-producing strains originated from Le Gatinais, Greneville-en-Beance (formerly Grigneville), Loiret and the Faronville districts of France. These bees were of varying and uncertain temperament, though highly thought of amongst the comb honey producers of the inter-war years. The late JM Ellis, of Gretna Green, boasted about two stocks of this type of bee which produced 240 heather sections. I recall the late Brother Adam having to beat a hasty retreat from bees of this type when they set about him. The late Colin Weightman and I made one visit to Buckfast Abbey to find Brother Adam taking action to overcome the onslaught. He had jumped into the mill race and was half submerged to avoid the bees' ire.

When moving to Yorkshire in 1972, I purchased some bees from the late Tom Brown of Harome, North Yorkshire, on the advice of Will Slinger and the late Beowulf Cooper, as these bees were 'heather bees'. Struck by their enthusiasm for such a good strain of bee, I took all ten stocks. The main (and most useful) characteristic of this strain of bee was that it built up very quickly and produced (as advised by Mr Brown) from five to eight queen cells from 4–8 May consistently year after year. These colonies were ideal for heather production;

they got swarming over early and settled down to build up for the late clover and the heather. They never lost this trait. Some years ago, I passed these colonies on to a commercial beekeeper when ill health forced me to reduce my stocks down to a modest half dozen. He told me last year that this strain was a real treasure for his heather honey production and for crops obtained from borage.

BROTHER ADAM, OBE, OSB

The late Brother Adam devoted almost a lifetime of 70 years to developing a new strain of honey bee, the Buckfast bee. He was named Karl Kehrle and born in Mittelbiberach, Germany, on 3 August 1898. As an 11-year-old boy of delicate health, in 1910, he arrived at Buckfast Abbey, a Benedictine monastery in Buckfastleigh, Devon, England, to devote himself to life as a monk.

In 1915, he began his work with the bees and, in 1919, he took over the responsibility for the abbey's apiaries from Brother Columban. In a very short time, he devoted his energy to building up the abbey's bee department. This included substantial construction works that saw the installation of very large ball-race bulk honey storage vessels and a steam operated heather press. All of this ran concurrently with the extensive honey production and bee breeding programme that he began in the early 1920s.

In the early 1960s, for three consecutive years, I spent a week at the abbey bee department enduring 12-hour working days with the bees. A day's activities were geared to work, with short ten-minute breaks in the morning and in the afternoon. To Brother Adam, the modern work-day

Strains of Bees for Honey Production

practice of a 30-minute lunch of pre-packed victuals was all that was necessary as it was deemed by him that a working lunch was all that was needed.

In one of my many visits to see him, Brother Adam told me that his ultimate aim was to establish a honey bee that would give beekeepers a constant average crop, consistent with minimum effort and time. Brother Adam's overriding principle of bee breeding was that none of the native races in themselves had all the qualities in a combination that he felt were required for good beekeeping practice in the UK. Brother Adam's ideas were to combine the best qualities from different races into a new super bee, possessing a combination of qualities that would give the modern beekeeper maximum surpluses with the minimum of work.

In its heyday, there were 320 honey-producing colonies at the abbey, plus 520 mating nuclei, the latter each on four combs. About 320–350 of these nuclei were usually overwintered and the young queens used for requeening the honey-producing colonies the following spring. Four to six colonies headed by a sister group were kept at the mating apiary at Sherberton, Dartmoor, for the purpose of drone production. Instrumental insemination was used for the cross-breeding experiments. The splendid isolation of the Sherberton apiary would ensure the use of drones of known origin with near certainty. Brother Adam's breeding work primarily consisted of two key groups. The first included the original strains of the Buckfast bee. The second was formed from his experimental crosses between other races and the Buckfast bee and their succeeding generations. The idea was to develop crossing into a line that would reveal reasonable genetic stability. Brother Adam reckoned on

Brother Adam's queen mating apiary, Sherberton, Princetown, Dartmoor, 1981 [Michael Badger]

(Left to right) Brother Adam with Phil Jenkin, Peter Donovan and Colin Weightman, exchanging ideas and thoughts at Brother Adams' inner sanctum at Buckfast Abbey, October 1980 [Michael Badger]

seven to eight years for this to be measured accurately. From about 1940 and again in the 1960s he developed combinations with the French black bee, and in the late 1950s to 1960 he persevered with a Greek (Carniolan type) bee that he used for incorporating traits from these races.

Brother Adam travelled the world over in his mission to perfect a super bee. Colin Weightman was his constant companion on many of these excursions, including his last trip to Turkey and the Mount Athos peninsula, Greece. Brother Adam retired from the work of the bee department on 2 February 1992, aged 93, living for a short while with his niece, Maria Kehrle, at his birthplace of Mittelbiberach before returning to the abbey until 1995. He ended his days at a nearby nursing home and died on 1 September 1996. My abiding recollection of Brother Adam was his vast reservoir of both mental and physical strength that never seemed to flag or diminish, despite rising daily at 4.45 am. He never seemed to tire, especially after a long day with the bees, after which he returned to monastic duties to round off his working day.

In one of my later visits to see Brother Adam, I asked him why he was so keen to breed from bees from other lands. His stock answer was that these breeds/strains very often possessed many advantages over our native stock, sufficient that this made it economically important to substitute the new for the old. However, in his book *Breeding the Honeybee*, he compares the virtues of different races and gives what he called *Mellifica lehzeni* (a western European mainland subspecies of *Apis mellifera*) very high scores on many of his most important measures, especially in relation to the Buckfast bee. His reason for backing the Buckfast over

A. m. mellifera in Britain was his belief that *A. m. mellifera* was extinct here. In this view he was wrong, as was found with the discovery of the Fountains Abbey bees by the late Bill Bielby. It not only never became extinct, but it can also be recreated by selective breeding of hybrids.

Over the years, I have met many people who denigrate Brother Adam's work on bee breeding; most of his detractors never met him, let alone discussed his methods or ideas with him. Yet today his work is valued. His strains of bees are being propagated in various parts of the world to good account.

SPURN POINT EXPERIMENTS

In the early 1970s, a number of Yorkshire beekeepers, under the enthusiastic lead of the late Bill Bielby, set up a breeding programme on the headland of Spurn Point, East Yorkshire. The late J Eric Hughes, Reg Spruce and Jack Renshaw actively visited the site. I recall visiting with Gerald Moxon and Bill Bielby on two separate occasions. The wind was a constant menace, with few bees flying, so much so that there were times when the bees needed feeding. I was not able to participate, other than at irregular intervals, as the site was a 200 mile round trip from my home and my work regularly took me away. After three years, evaluation found the conditions resulted in poor matings and the queens failed in a relatively short time.

A chance meeting with an oil rig superintendent at Phillips Petroleum, Easington, led to permission to place six hives with queen cells near to emergence on a redundant oil rig. The hives were carefully transported and landed in calm weather and the bees released. It was noted that the bees flew and drones were on the wing, too. After two weeks of moderate weather, the colonies were brought back to land. The queens had all emerged; they were found but had not mated. Some time later, we were advised that certain insects do not like flying over large expanses of water as it has the effect of both disorientating and chilling them. So it was a case of back to basics.

MINI NUCLEI FOR QUEEN MATING

With a little effort and some small investment in dedicated equipment, it is possible to improve the qualities of honey bee colonies significantly through queen rearing and develop personal beekeeping skills at the same time. There are many special techniques that may be employed and failure is a frequent occurence, especially at first attempts. Planning and scheduling each stage is vital. There are several techniques and commentaries to consider, but there are a lot of books and spoken words to help. A small team of three or four beekeepers working together helps enormously.

Some local beekeeping associations have experienced members willing to offer guidance on techniques. Glyn Davies from Ashburton, Devon, is very active with his local group. These are the points that Glyn emphasises with bee breeding work.

- First, select a breeder queen with the qualities desired and obtain healthy young larvae, less than 24 hours old, from within her colony.

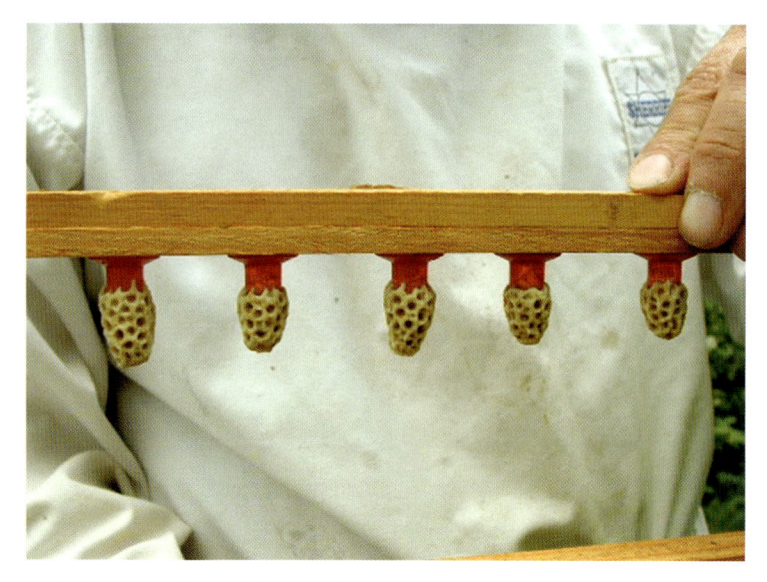

Grafted queen cells ready for transferring to mini nucleus hives [Glyn Davies]

- Second, prepare a good, healthy, large, gentle and, at least initially, queenless colony to raise the larvae to sealed queen cells.
- Third, identify a facility, aptly called a mating apiary, where your new queens can seek out their 11–15 or so drone partners and safely return home.

This is not the place to describe all the preparations that have to be made for each of these stages – there is plenty of literature and guides for consultation. Glyn is of the opinion there is nothing as good for learning as working within a group, especially if at least one person has some previous experience.

A mating apiary containing a number of mini nucleus hives [Glyn Davies]

A comb from a mini nucleus hive with a mated and laying queen [Glyn Davies]

FINAL THOUGHTS

Most beekeepers that I know have come to the consensus that some bees in their apiaries are better than others. There are strains that will work sections well, whereas others will not. It is up to the beekeeper to select those colonies best suited to the purposes for which he or she wishes to use them.

The trait that needs to be considered foremost is probably temperament. Over the years I have observed many strains of bees. The temperament varies a good deal. Some bees are born vicious, with the only option being to be rid of them. Unfortunately, bees of a vicious nature are very often regarded as good workers and tolerated for this reason; bees often produce better if they are disturbed less. Nature gave the honey bee a sting as a means of protecting the colony. It is a case of finding a happy medium. By careful selection it is possible to produce a strain of bee that may have excellent qualities of fecundity and docility and be prolific and industrious. These qualities make for successful beekeeping. The beekeeper can reduce incidents of stinging through careful manipulations and consideration of the following:

- regular washing of hands is necessary to remove the scent of bee venom as the associated pheromones excite the bees
- handle the bees in warm weather, avoiding cloudy thundery weather or windy conditions
- human odours and those of animals make the bees testy, possibly an inherited trait

Strains of Bees for Honey Production

- manipulations should not be undertaken when there is a dearth of nectar; bees are on their guard against intruders
- a colony without a queen is often testy. The absence of a queen is alien to the normal state of being
- robbing that is apparent. Small units of bees are always at the mercy of stronger colonies seeking a source of food
- the careful use of bee brushes; a goose wing is far better, but I prefer a handful of fresh grass.

Finally, the qualities we desire can be summarised as bees which:

- breed late into autumn and start early in spring
- store sufficient honey to overwinter
- overwinter well
- build neat comb and cap honey properly
- use the minimum of propolis
- are not 'followers'
- are of good temper, even when manipulated
- have a low swarming tendency and produce few queen cells
- have queens which are prolific and fecund
- are strong in constitution and longevity, disease resistant and show signs of hygienic behaviour
- are good foragers and will store large quantities of honey
- are not prone to robbing
- will defend themselves.

TYPES OF HIVE FOR WORKING HEATHER

It is recommended that single-walled types of hive are best for heather honey production because of their simplicity and ease of transportation to the moors. Those fortunate to live near heather moorland (close enough for the bees to be within flying distance) can use other hives such as the WBC double-walled type, or other types that are in vogue at the time. The use of polystyrene hives in preference to those made of wood is becoming increasingly commonplace.

PRINCIPLES OF BEE SPACE

It never ceases to amaze me how many beekeepers have a lack of understanding of the concept of 'bee space'. This is best described as the space that bees will tolerate within a hive without blocking or filling it with wax or propolis, thus providing free access to the various parts of the hive. Typically, this is in the range of 6–9 mm. The depth of either a brood chamber or a super is always one bee space greater than the depth of the frames to be contained. Hives are invariably described as 'top' or 'bottom' bee space.

Top Bee Space

Top bee space is a feature of Smith and Langstroth hives whereby there is a 6–9 mm space above the top of the frames in each box. This means the frames sitting within the brood chamber or super are set flush with the bottom of the chamber.

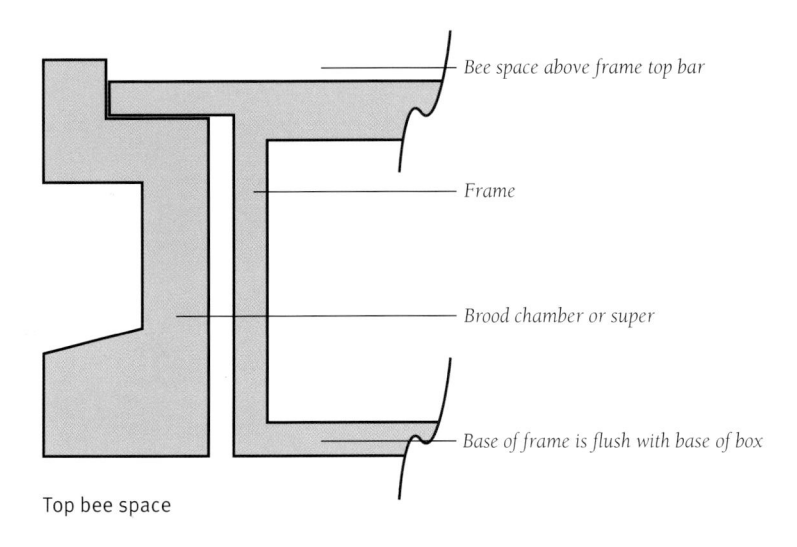

Types of Hive for Working Heather

Bee space above frame top bar

Frame

Brood chamber or super

Base of frame is flush with base of box

Top bee space

Bottom Bee Space

Bottom bee space is a characteristic of National, Modified National and Commercial hives, where the 6–9 mm space is below the bottom bars of the frames, with the frames being flush with the top of the chamber.

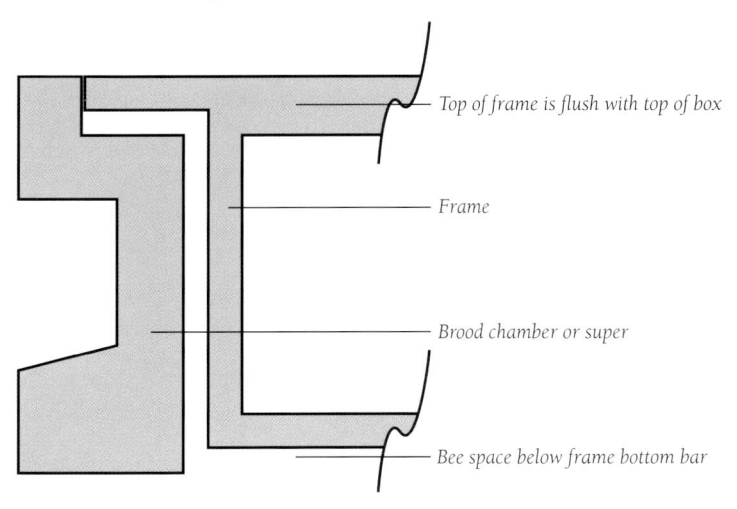

Top of frame is flush with top of box

Frame

Brood chamber or super

Bee space below frame bottom bar

Bottom bee space

151

During colony inspections, there is a key advantage for the beekeeper with top bee-space beehives (always providing the bee space is correctly sized). When prising apart two chambers of such a hive, the frames in the lower box do not lift as the chambers are parted. This is a known problem with bottom bee-space hives. The frames in the lower box are pulled upwards as the bees have often propolised the lugs of the frames to the box above. This inadvertent lifting of the frames has a tendency to upset the bees before the beekeeper really gets underway with the task of inspection. Therefore, there is a clear advantage in using top bee-space hives over bottom bee-space types.

The principle disadvantage with top bee-space hives is that you cannot place a chamber with bees directly onto a flat surface without sacrificing bees' lives, as they are invariably crushed should they be clinging to the underside of the frames. When using a bottom bee-space chamber, this does not readily occur, because of the gap retained.

STANDARDISATION OF BEEHIVE EQUIPMENT

The majority of hive manufacturers construct their timber hives and equipment to the now defunct British Standard (BS) 1300:1960. The collaboration of manufacturers over the past 50 years has ensured consistency, with standardisation allowing the different manufacturers' equipment to be fully interchangeable.

MATERIALS FOR HIVE MANUFACTURE

Canadian western red cedar has been the preferred timber for single-walled hives for over 120 years. It is light, durable, generally knot-free and relatively strong. It is unaffected by decay if the timber is from mature trees of a good age. In assembling the component parts, the use of waterproof glue is recommended. Stainless steel nails are preferred to galvanised nails, as the latter will rust away over time. There is a trend to use cheaper British western red cedar over Canadian imports. My experience of British western red cedar is that it appears to have an excessive number of knots.

The importation of redwood timber in the past three decades has made quite an impact as an alternative to expensive cedar wood. The use of non-toxic stains for preservation has made its use more attractive, despite the additional weight incurred.

There is a growing trend for use of other materials such as dense expanded polystyrene for hives for heather honey production as they are lighter in weight and have excellent insulation properties. The nature of the traditionalist heatherman was to use natural materials. Alas, progress deems otherwise.

FRAME TYPES

Types of frames and spacing methods can be a fairly emotive topic with some beekeepers, who often adopt a dogmatic stance. My view is that the beekeeper should use whatever he or she prefers. Beekeeping should be above dogma: it is what is best for you. The pros and cons are dealt with below. I have no prejudices on what should be used.

We know that the spacing of brood combs can vary but, on average, it is 35–37 mm from the midrib of one comb to the next. Drawn combs for honey storage can be spaced up to 50 mm from midrib to midrib.

Following the introduction of the movable-frame hive came the use of spacers fitted to the frame lugs or spacing manufactured into the frame design itself.

Three spacing methods were introduced for compliance with BS 1300:1960. These have continued in common use.

Metal or Plastic Frame Spacers

WBC metal or plastic spacers, colloquially referred to as 'metal ends', are slotted over the long lugs of the British Standard 355 mm x 216 mm frame to give a spacing of 35–37 mm between comb midribs. The use of metal ends (over plastic) can be tiresome because of their sharp edges which make them difficult to remove when they are propolised to the frame lugs. Frame spacers also have a disadvantage in that the combs are permitted to swing freely against one another when a box is being moved, which could result in bees and/or the queen being crushed. It is apparent that the lug widths of BS frames are fairly inconsistent in manufacture compared to 40 years ago, leading to loose or tight-fitting spacers. Metal ends are not suitable for use with short-lugged frames. The use of 11 National frames, with correctly spaced metal ends, leaves considerable space at one end of the chamber in which they are placed.

Hoffman Spacing

Hoffman frames are available for both brood chambers and supers, although I think that you pay a premium for super frames. It is not, in my opinion, money well spent for normal honey production. I do not advocate their use on the basis that the self-spacing varies from manufacturer to manufacturer. This inconsistency manifests itself with beekeepers finding they can use 12 frames over the norm of 11 frames in Modified National hives. Those who do use 12 frames will recognise that their colonies are apt to swarm more readily, perhaps owing to increased temperature which is self-generated through the increase in quantity of brood and the presence of extra bees themselves. This problem certainly manifests itself in very hot weather.

Hoffman frames in a brood box illustrating the spacing which creates provision for 12 frames within a Modified National brood box

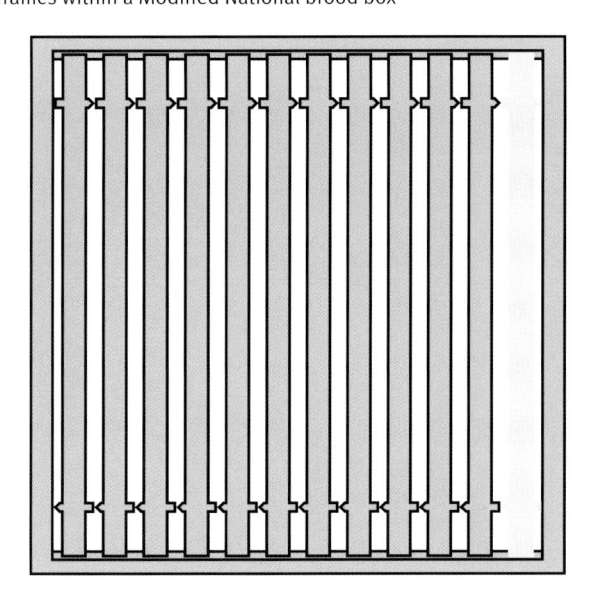

Manley Frames for Use in Honey Supers

Manley frames are favoured by serious beekeepers, whether commercial or large-scale hobbyists. They are rigid and robust and, as such, stay firmly in place. They are only available for supers. The spacing is 41 mm which ensures the minimum number of frames is used, yet avoids the danger of too wide a spacing; the bees are apt to build wild comb in the gaps and ignore foundation altogether. This is often the case if the foundation is not fresh or during a dearth or intermittent honey flow.

The main advantage of Manley frames is when you are uncapping. The design of the frame allows the operator to rest the uncapping knife on the top and bottom bars of a frame, which are of equal overall width. This allows for removal of the cappings with relative ease which, in turn, saves time. This is important for intensive labour operations. The Manley frame is ideal for both floral and heather honeys. Floral honey is best extracted with a radial extractor.

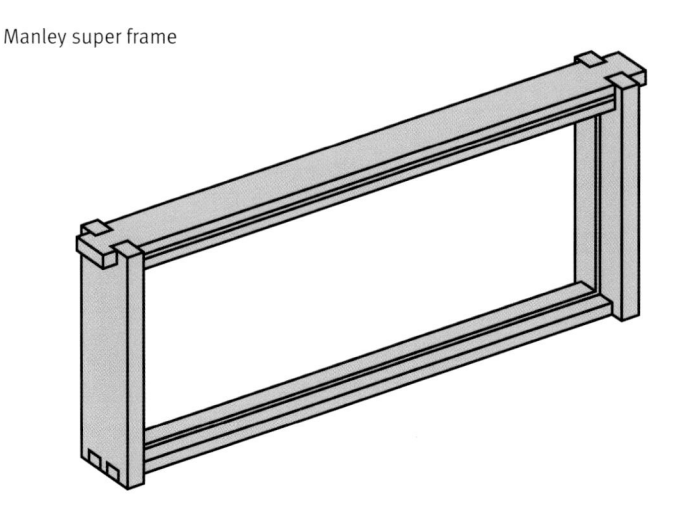

Manley super frame

Castellated Spacers

My preference for brood frame spacing is the use of castellated spacers. I prefer the fixed comb spacing and the method ensures the combs hang vertically and true, ensuring that foundation is drawn out neatly. Another feature is that the frames in supers do not move around so easily when travelling for migratory beekeeping purposes. Plastic or metal ends are not required, saving the labour of removing them for the process of honey extraction and reducing the potential risks of injury from sharp edges.

Those beekeepers hostile to the use of castellated spacing use the argument that they cannot push frames back and forth when undertaking manipulations. One could argue that pushing frames in this manner could quite easily crush or injure the queen. The main drawback can be that brood chambers with castellated spacing see a gradual build up of propolis on the base of the rebate over time, which will require removal at a subsequent spring clean. The simplest method to overcome this malady is to transfer the frames to a fresh brood chamber. The cleaning up is one of those winter day jobs or, as farmers would say, 'a wet day job'.

Castellated spacer

Propolisation

Different strains of bee use propolis to different extents. Propolis is a necessity as a means of sanitising the hive. If it were not so, the bees would not spend considerable time suffering great inconvenience collecting it.

Various styles of frame runner designed to reduce propolisation

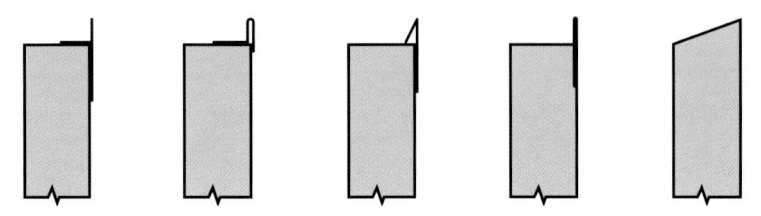

Short-lug and Long-lug Frames

In 1882, the British Beekeepers' Association fixed the sizes of both brood and super frames with long lugs. The much later introduction of new forms of beehive, with a rebate cut into the top edges of two opposite walls of the brood box made of only four pieces of wood, required the use of short lugs. This is a simpler type of hive construction and is used for Smith and Langstroth hives.

HIVE TYPES

The hives in common use today for migratory heather honey production are:

- British Standard Modified National or National
- Deep National

- Smith
- Modified Commercial
- Langstroth
- Hives made of expanded polystyrene

British Standard Modified National or National Hives

- Bottom bee-space design
- 11 combs
- Brood box: 460 mm x 460 mm x 225 mm
- Super box: 460 mm x 460 mm x 149 mm
- Long-lug brood frame: 432 mm (including 38 mm lugs) x 216 mm
- Long-lug super frame: 432 mm (including 38 mm lugs) x 140 mm

The Modified National hive is probably the most popular hive in use within the United Kingdom. It was developed by Arthur Abbott, Mountain Grey Apiaries, as an improvement to the original 'Burtt's of Gloucester', known as the 'Simplicity' or 'Commercial' hive. Following the Great War it was sponsored by the Ministry of Agriculture. Since the early 1930s the Simplicity or Commercial hive was superseded by the 'Modified' National type which was originally perfected in 1927. This modified hive has an added advantage over its other single-wall counterparts in that the top extensions provide excellent handholds compared with the finger recesses cut into walls of other designs, making lifting so much easier. These small handhold rebates were the major failing of the original National hive.

There is much to commend this type of hive, though there is much discussion about adequate availability of sufficient comb space for a moderately strong colony used for floral honey production. Additional space can be made available through the use of two brood boxes or a brood box plus a shallow box (normally used as a super). This does make colony inspections that much more difficult (finding the queen) but, on a positive note, it makes for less weight transmission when lifting parts of the hive.

Deep National

This hive is in every way the same as the Modified National hive except that the brood box accommodates frames 305 mm (12 in) deep compared with the Modified National which accommodates frames 216 mm (8½ in) deep. This means there are *circa* 80,000 available brood cells across the 11 brood frames compared with *circa* 54,000 cells in the Modified National. This hive is commonly referred to as a '14 x 12', reflecting the dimensions (in inches) of the frame.

Smith Hive

- Top bee-space design
- 11 combs
- Brood box: 466 mm x 416 mm x 225 mm
- Super box: 466 mm x 416 mm x 149 mm
- Short-lug brood frame: 394 mm (including 19 mm lugs) x 216 mm
- Short-lug super frame: 394 mm (including 19 mm lugs) x 140 mm

Types of Hive for Working Heather

The late Willie Smith of Innerleithen, Tweedale, was the inventor of the Smith hive; he was Scotland's first commercial beekeeper. It is a hive of simple design that uses short-lugged British Standard frames. The Smith hive is based upon the single-walled hives of the United States of America. The internal capacity is identical to that of the National hive. The combs hang at right angles to the hive entrance. It is a top bee-space arrangement whereas the Modified National hive is bottom bee-space. The Smith hive's Scottish origins make it a very popular hive with beekeepers in Scotland and the border counties.

Modified Commercial Hive

- Bottom bee-space design
- 11 combs
- Brood box: 465 mm x 465 mm x 266 mm
- Super box: 465 mm x 465 mm x 162 mm
- Short-lug brood frame: 438 mm (including 19 mm lugs) x 254 mm
- Short-lug super frame: 438 mm (including 19 mm lugs) x 152 mm

The brood box and supers are of a simple design and construction. The design is ideal for those beekeepers wishing to use a prolific queen contained in a single brood chamber. The floor, roof and crownboard are more or less identical to those of the Modified National hive so that each is interchangeable.

A major disadvantage is its weight when fully laden, especially for those amongst us who are not particularly strong or are incapable of heavy lifting.

Langstroth Hive

- Top bee-space design
- 10 combs
- Brood box: 508 mm x 413 mm x 240 mm
- Super box: 508 mm x 413 mm x 146 mm
- Short-lug brood frame: 483 mm (including 19 mm lugs) x 232 mm
- Short-lug super frame: 483 mm (including 19 mm lugs) x 136 mm

The Langstroth hive is the most extensively used beehive the world over. It is of American design. However, other than with a minority, it has not become popular with beekeepers in the British Isles. The brood chamber, according to Dr Colin Butler's MAFF Bulletin No 144, *Bee Hives*, is made to accommodate ten frames, giving a maximum total comb area of 2740 sq in. This is around 500 sq in more than that of the 11 comparatively smaller brood frames used for the Modified National or Smith brood chamber. When using a Langstroth hive, it is not unusual for beekeepers to add another brood or super chamber to accommodate the brood of a prolific queen. Given this point, the Langstroth hive has little to commend it over its British counterparts. Perhaps it is for this particular reason that it has lacked appeal to many British beekeepers for at least 120 years. A major disadvantage is its weight when fully laden.

Types of Hive for Working Heather

POLYSTYRENE HIVES

The manufacturers and dealers of expanded polystyrene hives suggest that they have many benefits over their wooden counterparts. Without doubt, they are lighter and possibly cheaper, with excellent insulation properties that assist the bees to overwinter and promote a more rapid spring build up of colonies. In addition to flat packs, hive components are available in one-piece mouldings, so no assembly is required. A good coating of proprietary masonry paint to the outer surfaces of the hive is recommended to reduce the effects of natural ultra-violet light that otherwise may cause degradation of the material. The polystyrene has at least four times higher density than typical packaging polystyrene, making it robust. Properly cared for, these products have a useful life expectancy of more than 30 years.

It is advisable to protect polystyrene hives from rats and woodpeckers using 15 mm chicken wire to encapsulate the hives and protect them from damage by predators, especially in severe winter weather.

Hives are available in a range of sizes compatible with standard size frames, including National, Deep National and Langstroth. All of the main components – floors (integral or separate), brood boxes, supers, feeders, roofs – are usually constructed of polystyrene. Ancillary components such as queen excluders and crownboards are usually made of plastic.

OTHER HIVE TYPES

Without wishing to disparage other types of hive that are available – top-bar hives, the Warré hive, the Dartington hive, the catenary hive, etc – these are really not suitable for migratory beekeeping. The most suitable hives are simple and compact. Traditional tried and tested methods of hive management are most suitable for heather honey production.

OTHER CONSIDERATIONS

It should be remembered that Commercial, Deep National and Langstroth hives are heavy and cumbersome to lift when loaded with bees and honey, especially when the hives require moving or transporting, whether loaded with honey or not. It is a feature that cannot be easily overlooked for heather honey production, so:

- whichever hive is chosen, it needs to be compact, with close-fitting boxes
- it should be covered with a shallow flat roof
- the top of the super should be covered with 25 mm thick insulation board when the hive is on the moor; good packing will never go amiss
- a ventilation screen should always be used when moving a hive, especially if the weather is warm
- the hive entrance should be completely covered and secured by an entrance block or foam strip during transit to and from the moor
- perforated zinc should not be used to cover the hive entrance. It creates excitement for the bees trying to escape to the daylight. This raises the temperature which, in turn, reduces the bees' life – if it does not kill them

Types of Hive for Working Heather

A hive with eke, properly prepared for moving with double straps, a ventilation screen and foam strip to block the entrance [Michael Badger]

- blocking off the entrance should be done immediately prior to departure. It is ill-advised to shut bees up prematurely as a strong colony will be ruined through lack of ventilation; better to lose some flying bees than risk permanent damage to a colony, especially on a warm evening as can be expected in early August.

The late Harry Grainger of Leeds was not a physically strong individual, so he made up strong colonies by uniting swarms into shallow supers restricted to nine frames with the use of dummy boards. Because of their small size, he could lift and carry them with ease to the back seat of his car. All the years that I knew Harry, he always returned from the heather each year with a good surplus.

OTHER HIVES FOR HEATHER HONEY PRODUCTION FROM DAYS GONE BY

Baker Hive

The Baker hive was named after its creator, a Yorkshire commercial beekeeper who ran a large enterprise in Pickering, North Yorkshire. He developed a combined floor and brood box with a lockable entrance for migratory beekeeping. The brood box was designed to hold ten British Standard frames; the dimensions permitted the upper chambers to take National size frames. The hive was heavy and cumbersome and required two people to lift it. It was made from redwood, painted externally with white lead. The roof was of the same design as the WBC.

Baker hives at Arthur Scaife's apiary, Pickering, North Yorkshire, 1959.
(Left) Arthur Scaife, (right) Sammy Henderson [William Slinger]

Glen Imperial Hive, No 4

From its external appearance, the Glen hive looks very much like a WBC hive, with external lifts. It was constructed for heather honey production, holding 15 BS frames, although its weight makes it only suitable for non-migratory beekeeping. Its use would have been confined to stances that bordered the heather. Its entrance is of a full-width funnel design. The queen excluder is framed with intermediate supports to prevent sagging.

Congested Districts Board (CDB) Hive

This hive was designed on similar principles to the WBC hive but it holds 11 BS frames. It originated at the time of the Land Law (Ireland) Act 1881, for tenant farmers in Ireland. Sixteen years later, the Act was extended to Scotland under the direction of the Congested Districts Board (Scotland). The

Glen Imperial hive. Steele & Brodie catalogue, 1934

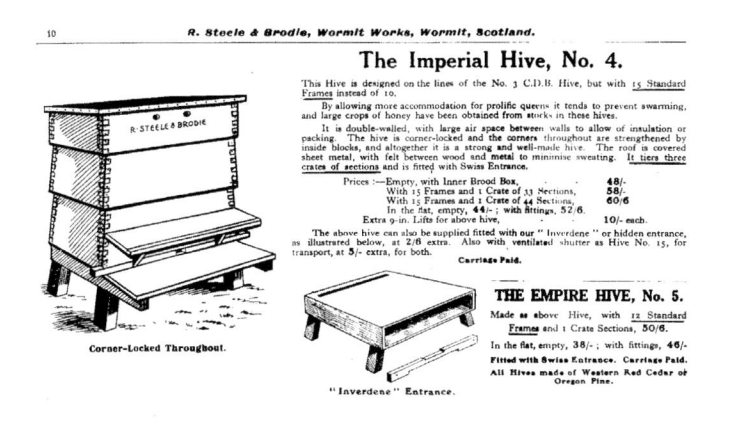

Congested Districts Board hive. Steele & Brodie catalogue, 1934

introduction of these hives was to support cottage industries given finance for carpet making and lace making and was extended to beekeeping with funds for both equipment and beekeeping instruction. The hives were made by both Steele & Brodie, Dundee, and Abbott Bros, Dublin.

The Author's Migratory Heather Floor Board

This type of floor is ideal for heather use only. The hinged entrance will quickly become propolised if used all year round, unless the hinges for the drop-down entrance are replaced with thick leather or some synthetic pliable material to give the same flexibility. These floors are robust and are ideal for migratory heather beekeeping as they make it quick to close the entrance.

Types of Hive for Working Heather

The Author's migratory heather floor board [Damien Timms]

THE IMPORTANCE OF GOOD QUALITY DRAWN COMBS

There were honey bee colonies at my school which were tended by me and my chums. Our music teacher provided encouragement and support, also supplemented by regular visits from Harry Allen, Warwickshire County Beekeeping Instructor, better known to my generation as the 'Combings' correspondent of the now defunct *British Bee Journal*. Quite by chance, I came across my school beekeeping logbook that had an interesting item concerning the value of keeping good honey combs for immediate and future use in the apiary.

While my log book was written more than 60 years ago, the information given then is equally relevant to beekeepers now. We were told that quality drawn combs are of great value to the beekeeper, more so than good hives. We are all too aware that any sort of hive will serve as a shelter for the bees. The bees' choice of home at swarming time testifies to this as it can be anything from a chimney stack to a church steeple. Therefore, it is safe to say that the type of hive has no real effect on the prosperity of the colony, always provided that it fulfils certain basic conditions. In their wild state, the bees' natural choice of nest site is generally high up above the ground. It is beekeepers who keep bees at low level for the benefit of practicality. All that the bees require is a sound, dry, predator-proof home. Under such conditions, as long as the colony has space to expand, all will be well. It is the beekeeper who fancies a particular hive type or a hive design that satisfies his or her preferences. It is not the choice of his or her bees.

Modern hive management as we know it today is based upon four basic principles:

- the existence of bee space
- the elasticity of the compartment created by the beehive
- the building of combs in frames which are removable and interchangeable
- disease control and the issue of dealing with exotic pests.

One could say that none of these principles should frustrate the normal instinct of the bee colony and should always be in harmony with it. I have found that those new beekeepers who run into trouble do so because their management techniques run contrary to the needs of the colony. Good beekeeping in all its aspects results not so much from an encyclopaedic knowledge of operational techniques, but from close study of the ways of the bees themselves and of the patterns of colony behaviour seen in their wild existence.

Newly Produced Honey Storage Combs

If they are looked after with reasonable care, which they normally get, combs that are used exclusively for honey collection and its storage prior to extraction will stay in good condition almost indefinitely. Care is needed if using a tangential extractor for extracting honey from newly drawn combs as the wax is fragile and can be so easily damaged as a result of centrifugal force exerted at right angles to the face of the comb. Combs extracted using a radial extractor are less likely to be affected in the same way as the dynamic force acts in parallel to the faces of the comb.

Tangential action (left) and radial action (right). Rotation is in a clockwise direction. The centrifugal force acting on the frame is indicated in each case

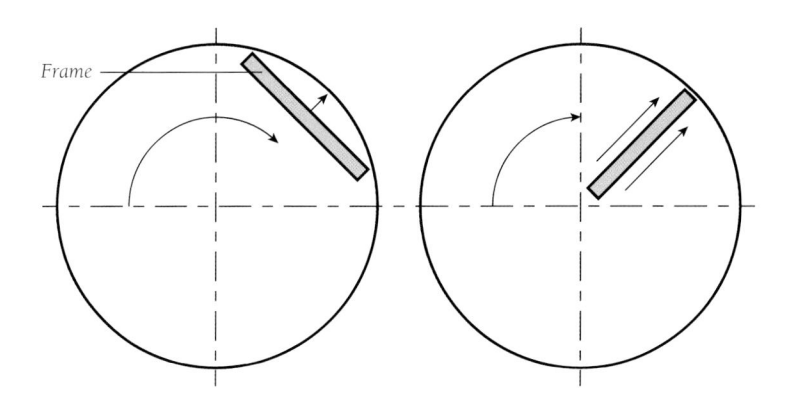

It is suffice to say the beeswax of honey combs from supers remains light in colour, it is generally clean and, if combs are used without loss as a consequence of oilseed rape (OSR) honey crystallising in the cells, annual uncapping results in the regular addition of new wax.

Beekeeping in urban areas has an advantage over beekeeping in the countryside, where foraging colonies may work OSR, in terms of retaining good comb for future use. Those combs unaffected by OSR rarely remain on a hive long enough to become badly stained by propolis or through dysentery. Furthermore, brood cocoons and pollen are absent. The presence of a honey film inside the cells of such combs discourages the activities of wax moths and pollen mites. These honey super combs are easy to store over winter.

Replacement of Brood Combs

Brood combs are in a totally different category from honey storage combs. They are apt to deteriorate rapidly and their useful life is so much less than many beekeepers fancy.

In the wild state, honey bees do not continue to use the same comb as brood space forever. Given the opportunity, a honey bee colony will extend its brood area downwards each year, gradually working from old comb to new. Bees seem to like to overwinter on old comb, preferring new comb for early spring development and summer use.

Beekeepers who understand this phenomenon will allow each colony to produce some brood comb each year and will introduce frames of foundation to be drawn on the flanks of the brood nest (but not next to the hive walls) so that the brood area can expand naturally. Never place frames of foundation

Types of Hive for Working Heather

into the heart of the brood nest in early spring as this creates a division, which may result in the production of emergency queen cells. William Hamilton, in *The Art of Beekeeping*, states: 'The provision of new foundation acts as a tonic to an expanding colony'. It works by encouraging normal and natural development of the colony and it is also one of the natural measures which prevents or delays swarming.

Over a period of two years use, brood comb loses 25 per cent of its available space in which to develop worker bees because the bees will remodel the comb over time, for example by creating drone brood cells. This has a significant impact on the build up of colonies, so essential in the early spring, to provide the number of foragers needed to maximise honey production.

As the beekeeper adds new foundation, old and poor combs can be discarded by working such combs to the flanks of the brood chamber, thereby discouraging their continued use by the queen. Such combs may be removed once the brood they contain has emerged. Combs that are found to be worthy of another year's use (always providing they are disease free and have no dysenteric staining) can be used for nuclei or swarms can be hived on them while the bees draw out foundation in frames placed in a brood chamber above. Once the queen is laying in the new comb, a queen excluder can be inserted between the upper and lower chambers to prevent further access to the old combs which, in turn, can be discarded once the brood has emerged.

Combs of drone brood may be scrapped out of hand, but if these combs are also free of disease and staining, they may be used to good effect in the apiary for early queen-raising operations. It is not always wise to destine them for the solar wax extractor.

Colonies of honey bees are most harmonious when drones are present. They are all part of colony cohesion and a vital aspect a colony's existence. Total elimination of drones and, therefore, comb in which to rear them, is not desirable. That said, an excess of drones in a honey-producing colony is not an advantage when surplus honey is required for harvesting.

I have found that damaged worker brood combs can be revitalised by scraping them down to the midrib and inserting them into a nucleus colony that is building up in strength, always providing the combs are from a disease-free colony and do not exhibit dysenteric staining. The nucleus wants worker bees so the rebuilding of comb with worker cells comes about through necessity.

Poorly-shaped and Incomplete Combs

Poorly-shaped and incomplete combs are obvious choices for discarding and removal in most situations and circumstances. They may be eliminated as a matter of routine. However, frames of noticeably newish comb can be salvaged by carefully scraping it down to the midrib and giving the frames to a nucleus colony as described above.

It is also possible to give such frames to a newly housed swarm which will work to build comb with alacrity.

Very Dark Combs

Such combs, even if they appear to be in good shape, should not be retained in use for too long. Darkened comb is a natural sign of age. As a comb gets older the cells become smaller because of the micro build up of the larval cocoon skins of successive generations of bees. As the cells become smaller, so do the bees produced, and this can be quite noticeable. Even if some dimunition in size is not apparent, it is nontheless there. These smaller bees are likely to be deficient in other physical respects and will be less useful as workers, a fact that is so often overlooked.

Every colony of bees should be encouraged to draw at least two sheets of foundation into brood comb each year, when the colony is expanding and when it would normally be building new comb. This will depend on the colony strength and whether the season is early or late. Two sheets of foundation may be inserted at the same time provided that one frame of foundation is placed on each flank of the brood nest. The ideal situation is between a pollen comb and an adjacent frame of eggs and larvae. Introducing one frame of

A well-drawn brood comb. Note the wax comb extends to all edges of the frame, providing the maximum area of brood cells [Damien Timms]

foundation at a time is a better plan. It is not recommended that two sheets of foundation should be placed side by side unless they are given to a swarm, or if a shortage of drawn combs makes this unavoidable. In the latter case, the outer frame of foundation should be moved one space inward as soon as the first frame of foundation is drawn and occupied with eggs and larvae.

The best produced combs are drawn when placed in an upper brood chamber. In this position the foundation gets drawn to the bottom bars of the frames. This rarely, if ever, happens when frames with foundation are drawn in a single or lower brood chamber placed directly onto the hive floor.

There are exceptions to this, such as with the use of a special gadget known as a Killion board. This is an apparatus developed in the USA by Carl E Killion Comb Honey Producers in the early part of the 1920s. In plan, this board looks in every respect like a slatted showering board. The board is placed between the floor of the hive and the brood box. The construction and design of this special floor permits the bees to cluster below the bottom bars of the frames, enabling them to maintain the temperature required for drawing wax neatly down to the bottom bars.

The normal cross-sectional arrangement of a brood nest is: honey, pollen, brood (eggs, larvae and pupae in all stages of development), pollen, honey. Bees expand the brood nest outwards by consuming the pollen from the centre of the pollen combs. Nurse bees then prepare the emptied cells for the queen to lay in.

When using National, Smith or WBC hives, which have relatively small single brood chambers, it will be observed that the frames in the super placed immediately above the queen

Types of Hive for Working Heather

A Killion board. Note there is a 25 mm space both above and below the slats, to provide sufficient space when it is placed on the floor board [Michael Badger]

excluder may contain arc-shaped layers of pollen and honey, colloquially known as a 'pollen arch'. Those beekeepers who work with double brood chambers will notice this on frames contained in the top brood box. Beekeepers using brood boxes of greater depth – Commercials or Deep Nationals, for example – will observe this on the individual brood combs. The best place for frames of foundation is between a pollen comb and the adjacent frame of brood. No other place or position is as good. The only time when this cannot be done is when all the combs within the brood chamber contain brood in an advanced stage of development. Never insert frames of foundation in the outermost positions in the brood chamber. This is the area where drones are known to congregate. Foundation in this position cannot be properly drawn and, frequently, only one face of the foundation will be tackled by the wax-producing bees. This is a most unnatural situation since bees invariably work on both faces of a comb at the same time. As stated previously, under no circumstances should frames of foundation be placed in the heart of the brood nest.

This general rule applies also to honey storage supers. Keep frames of foundation to the outside of existing combs unless there is a heavy nectar flow. Avoid having foundation in the end frame in any box (brood or super) at any time. A box of foundation with a drawn comb on each flank is a good starting point. A mixed super of comb and foundation should be arranged so a drawn comb is on each flank. The spacing is critical, too. Do not position frames of foundation on wide spacing as the bees will generally opt for producing burr comb between adjacent faces of foundation as the distance between them is wider than the natural spacing of comb. A good method is to reduce spacing and insert an extra frame (12 frames in a National or Smith super or 11 frames in a WBC). Once the bees have started to draw the foundation in this additional frame, it can be removed. Once the combs are fully drawn (and usually the bees are starting to fill them with nectar) they can be spaced more widely. A mixed super of drawn comb and frames of foundation should be arranged so that a comb is on each flank, the rest of the drawn combs in the centre, with frames of foundation used to fill the remaining spaces. It should be noted that if unfavourable weather intervenes, the bees often defy logic and produce brace comb between the foundation in adjacent frames. This is also often the case if beekeepers use 'stale' foundation or place the frames with foundation at wider spacings.

Generally, a super of foundation should only be placed above a drawn super until the combs in the centre have been drawn. Two of these combs should then be moved to the flanks so that frames of foundation never come next to the external sides of the super box. Again, in oilseed rape nectar flows, care needs to be taken because ideal weather conditions can change within hours to cold and wet. The novice has to come to terms with this.

A queen excluder should be placed between the brood box and the super only when the bees are occupying the super. Should the queen begin laying in the super, drive her down first with a puff of smoke or, better still, pick her up and place her down below to ensure she is in the brood nest and beneath the excluder.

The preceding narrative is a generalisation for good comb management practice, essential to good bee husbandry. Comb management for the two heather crops is discussed elsewhere, under the section on production methods.

Some Considerations for Comb-building Nucleus Stocks

The use of nuclei to produce drawn combs requires a considerable amount of skill on the part of the beekeeper. The less experienced beekeeper can so easily fail in his or her endeavours as building up a small stock in this way is at odds with the natural development of the colony. This is a prime consideration that is so often overlooked in this form of colony management.

A colony, with its growing population, must *never* suffer a check in its development. It beholds the beekeeper to ensure that he or she avoids such a situation, as this will trigger the production of swarm cells. At all times, the beekeeper has to

The Author's nucleus hive comb-building stocks [Hilary Badger]

Types of Hive for Working Heather

be ahead of the bees by checking the amount of brood being produced. The 'chimney effect' that is created with this method of production forces the colony into an unnatural situation. That said, stray swarms often house themselves in restrictive conditions and subsequently swarm on an annual basis. This situation brings both a rise in temperature and congestion. These two factors most often trigger swarming preparations. At all times there must be plenty of room to dissipate heat that rises naturally and this must be allowed to escape. A simple solution is to place four matchsticks, one at each corner, immediately below the crownboard to create a gap which which lets out the excess heat. I have found that it is only later in the season that the bees propolise this gap. Place a couple of deeper boxes above the crownboard, complete with ready made-up frames of foundation. This allows the bees some refuge and gives some room, allowing heat to dissipate from the heart of the brood nest.

To keep the colonies small, the simplest solution is to remove sealed combs of brood and distribute them among weaker colonies. With disease-free stocks, such action is not to be frowned upon. Should you find that your main stocks do not need reinforcement, such combs (that must contain only sealed brood; no eggs or larvae must be present) can be placed over the crownboard of an established stock. The brood will emerge in due course. If any combs with eggs and larvae are separated in this way, it will trigger production of emergency queen cells and the whole plan goes haywire. Use an empty super as an eke and place an empty frame with no foundation within it flat on the crownboard. Place the frame of brood on top of this and then place another similarly empty frame on top. Cover and encapsulate the arrangment to retain heat. This will allow the colony's nurse bees to nurture this brood until it emerges.

The work entailed for comb production in nuclei may not be for the faint-hearted, but one could say that nothing ventured, nothing gained. This method is fairly labour intensive, the nuclei requiring daily attention to ensure success. It is a system that can easily fail without this commitment.

SYSTEMS OF MANAGEMENT FOR BELL HEATHER HONEY PRODUCTION

Colonies to be used for harvesting bell heather honey generally require similar management and stockmanship to those to be used for other summer honey production.

However, a number of simple amendments need to be made for bell heather honey production. The prime consideration is that the colonies *must* be exceedingly strong to work bell heather and cope with the varying climatic conditions experienced in the British Isles.

There are a number of methods available that can be used for bell heather honey production, or any other crop that comes late such as clover, evening primrose, phacelia or borage. The following two methods have been found to meet the needs of both new and experienced beekeepers:

- a method of 'doubling' for honey production
- the Demaree system of controlling swarming and for honey production.

THE DOUBLING METHOD

The doubling method has been found to be a good method for bell heather honey production as this honey flow is relatively late; it can commence from mid to late June or even later, naturally dependent on latitude, weather and season. While this system of management is ideal for National, Smith and WBC hives, it cannot be recommended for those hive types with deeper brood boxes. This method was a favourite with both Mountain Grey Apiaries and Yorkshire Apiaries of East Yorkshire. These two large commercial concerns worked the Yorkshire and Lincolnshire Wolds for Kentish white clover and the North Yorkshire moors for bell heather from the early 1930s until the 1960s with considerable success.

The primary objective is to rear as much brood as possible before mid June and then reduce brood production. This maximises the field force of foraging bees to collect nectar for surplus rather than directing colony resources to the feeding of developing brood. It is essential that the brood combs are not old or clogged with stale pollen that is hard and unusable and that there are no more than two combs of sealed stores in the brood box. It must be remembered that brood combs lose 25 per cent of their usable area for brood rearing after two seasons.

Mention is made later of the merits of overwintering colonies on two brood chambers for National, Smith and WBC hive types. Colonies that have been wintered on two brood chambers should not usually be disturbed until early to mid April, depending upon the season. Again, this will vary depending on locality.

This system is particularly suitable for urban beekeeping or other locations where oilseed rape is not grown. Around the beginning of April, or earlier if the season is early, brood will be found in the upper brood chamber. A quick inspection should be made to determine the extent of the brood. In doing this, the brood nest should not be disturbed. Removing frames of brood comb may set a colony back. It should be sufficient to gauge the extent of the brood by removing the crownboard and peering between the frame top bars.

Carefully take off the upper box and set it to one side. Look through the lower box of combs to determine the extent of any brood. If these combs contain brood, replace the upper brood box and leave the colony as it was found. However, if no brood is found in these combs then the position of the two brood chambers should be reversed. That is, the original upper box is placed on the floor board with the other placed on top. This is also an ideal opportunity to replace the existing floor board with a clean one. Should it be apparent that surplus nectar is being brought in, a super box may also be added.

Colonies overwintered in single brood chambers with bees covering eight combs or more are ready for doubling. The additional chamber can either be a deep or a shallow chamber, preferably with frames of drawn comb. Cover the arrangement with insulation board or sacking on top of the crownboard. The colony should be well provisioned with sealed stores or fed with syrup.

If frames of drawn comb are not available, frames containing foundation will have to suffice. Two combs of brood plus the attendant bees from the existing brood chamber should be placed in the centre of the new brood box. Rearrange the combs of brood in the exisiting brood box so that new frames of foundation can be placed next to the pollen comb on both flanks of the brood, as mentioned previously. The new brood chamber containing the two combs of brood should be filled with frames of foundation and placed on top of the existing brood box. Place a feeder on top to assist the bees to draw out the foundation. Feeding may not be necessary if the weather is kind and nectar is being collected. The colony should be left alone for at least two weeks but observation of the activity at the hive entrance will tell you much.

It is a natural phenomenon that all swarms start new comb from the top of their cavity. Bees invariably work downwards, provided there is no space above them. Placing a new brood box on top of the existing brood box is contrary to the way bees live in nature and encourages the bees to fill this added space at the earliest opportunity. With incoming nectar, or syrup, the colony will draw new comb at will.

The heat rising from the cluster makes it easier for queen and bees alike to work. Understandably, heat is lost, no doubt no more than would be necessary to raise additional brood in the lower brood chamber. The combined effect of heat above the cluster and the natural urge to fill the vacant space with brood makes the colony develop faster. If carried out correctly with insulation above the crownboard, there will be no chilling of brood.

Possible Risk of the Doubling Method

The only risk with the doubling system is the reliance on favourable weather from mid April until early May. Should the weather be cold and wet, a considerable amount of syrup feeding may well be required. It goes without saying that this system involves a great deal of food consumption by the bees compared with other methods. The end results make it well worthwhile.

Next Steps

Depending upon the conditions, the queen excluder should be placed between the two brood chambers no later than mid June, ensuring the queen is in the lower of the two chambers.

In early seasons this action may be carried out early in the month. (Note: irrespective of the earliness or lateness of the season, the hours of daylight are related to the calendar, eg, equal hours of night and day at the summer solstice, 21 June.) There are three reasons for carrying out this action.

- First, the rearing of brood needs to be curtailed as there is a risk of over-production, whereby incoming nectar will be diverted to support additional brood rather than stored as honey in the supers. In addition, thousands of bees would be mobilised to undertake nursing of brood rather than being out in the field foraging for nectar.
- Second, the nurse bees, along with the queen, have a tendency to neglect the combs in the bottom brood box. The introduction of the queen excluder keeps the queen in a position where the urge to lay soon abates.
- Third, the tendency to swarm is reduced. The introduction of the queen excluder creates an unnatural unoccupied space between the brood and the stored honey within the supers,

SWARM CONTROL AND HEATHER HONEY PRODUCTION

The problem of swarming must not be overlooked when stocks are worked for bell heather. Colonies will be 'home' to their own drones, a prerequisite to the inducement of swarming preparations, as mentioned previously.

Swarming preparations can start whenever conditions warrant it and on the moor is no exception. Colonies can swarm at any time when they have their own drones on the

A swarm that decamped onto a rock face at Dalby Moor, North York Moors National Park, and survived until January the following year [Brian Nellist]

wing, a point that many experienced beekeepers frequently overlook. It needs to be understood that from the earliest point in the season, swarming can happen once a colony's drones are flying; it is the surest sign for the beekeeper of the start of the swarming season. Drones may be present right up to the end of July but, usually, swarms are rare at this time.

The large field force of foragers brings the issue of swarm control into play. In addition to having a young queen, sufficient room, ventilation and plenty of good combs, the beekeeper needs to be ahead of the game in controlling the bees' natural urge to swarm – a situation that is bound to arise with a colony bursting with foraging bees. The main secret, if there is one, is to ensure that at no time is is there a check in the development of the colony. This is most important. Remember that no swarm will issue without its queen and also that a colony with a clipped queen may still attempt to swarm.

A number of swarm control methods have much to commend them. These include the Pagden method supplemented with the Heddon method of reinforcement, the Demaree methods, or the use of the Snelgrove or Horsley board. The beekeepers of old used the flowering of the currant or red-flowering currant (*Ribes sanguineum*) as an indicator of when they should do their first brood nest inspections to check for swarm cells. Climate change has led to early-season development of colonies that is far more advanced than would have been expected 50 years ago. It behoves all beekeepers to become observers of the hive entrance; such observations will tell you much about the colony without the need to open up the hive and disturb the brood nest.

Pagden Method Supplemented with the Heddon Method of Reinforcement

The new beekeeper can do no more than become proficient in the use of an artificial swarm method to control swarms until he or she feels more confident to use other systems. The best method for the novice is the Pagden method, named after its originator JW Pagden, supplemented with the Heddon method of reinforcement. The Pagden method was originally published in 1870 and used for fixed-comb skep hives. It worked equally well for the movable comb hive. With this method it has to be considered that while swarming will be prevented by taking the brood away, leaving only flying bees and the queen on empty combs or plain foundation, there is a cost in terms of a reduction in surplus honey. With the need to start brood production, the whole mass of foraging bees will naturally be distracted or diverted from the supers to the

Artificial swarm

1 Remove the parent stock to a new site at least 2 m (6 ft) from its original position. The entrance of the original hive on the new site is turned at 90° to its original orientation

Queen
Queen cells

2 Place a new floor and brood chamber on the original site and put the queen from the parent stock, along with one frame of sealed brood with no queen cells, in the new brood chamber. Fill the remaining space in this brood chamber with frames of drawn comb or foundation

Queen plus one frame of sealed brood with no queen cells

Queen cells

3 Take the supers and queen excluder from the parent colony and place them on top of the new brood chamber, containing the queen, which stands on the original site.

Queen plus one frame of sealed brood with no queen cells

Queen cells

4 Close up both hives

Queen plus one frame of sealed brood with no queen cells

Queen cells

Heddon reinforcement

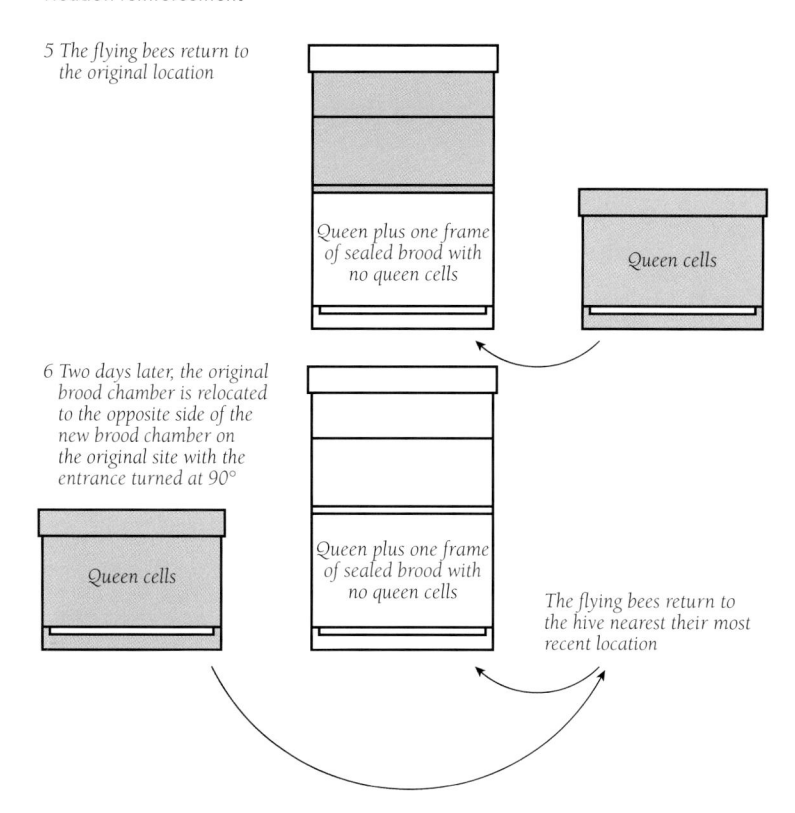

5 The flying bees return to the original location

Queen plus one frame of sealed brood with no queen cells

Queen cells

6 Two days later, the original brood chamber is relocated to the opposite side of the new brood chamber on the original site with the entrance turned at 90°

Queen cells

Queen plus one frame of sealed brood with no queen cells

The flying bees return to the hive nearest their most recent location

brood chamber, unless the honey flow should be exceptional and sustained. The Heddon method of reinforcement was devised by a beekeeper of the same name from Devon. The reinforcement of the number of flying bees makes the Pagden method that much more successful.

The principle is simplicity at its best. The 'swarm' (consisting of the queen and the flying bees) is hived in a new brood chamber and set on the stand previously occupied by the parent stock. The honey supers are added over this new box. The parent stock is relocated adjacent to the artificial swarm. It is suggested that it is positioned at right angles (if space permits) to its original orientation. This action permits the artificial swarm to be reinforced by all the flying bees, which naturally return to the location of the original hive.

After two days, the parent stock is then moved to the opposite side of the hive containing the artificial swarm, once again at right angles to its original orientation. This action further depletes the parent colony of its flying bees as they return to the hive nearest this colony's most recent location (ie, the artificial swarm). After a further two days, the parent stock is moved carefully to another part of the apiary, further reducing the number of flying bees. These manipulations have reduced the number flying bees to such an extent that the parent stock is unable to throw a cast. There is no need for the beekeeper to destroy queen cells as the emerging queens will naturally resolve the matter for the beekeeper.

Should a beekeeper have a use for surplus queen cells, these can be removed from the parent stock prior to emergence of the queens, or the bees can be helped by selecting a queen cell in a good position and removing the remainder, thus avoiding the queens fighting it out as nature intended with possible injury to them during the skirmish.

The object of these operations is twofold: the flying bees are compelled to drift into the hive on the original site, thereby supplementing the number of foragers; the series of moves depletes the parent stock of its flying bees and any casts or after-swarms are avoided as the colony is not in a balanced state.

Method

1. The queen is taken, along with a frame of sealed brood without any queen cells, and placed into a new brood chamber on a new floor. This is then filled with drawn combs or, failing that, with frames of foundation.
2. The new hive containing the queen (artificial swarm) is placed on the site of the original hive (parent stock). A queen excluder in good condition is placed on top.
3. The supers and bees therein from the parent stock are placed onto the hive containing the queen. A crownboard and roof are added.
4. The parent stock is now reduced to a floor board plus the original brood chamber with combs of brood and queen cells. An empty comb is placed at one flank to fill the brood box and a crownboard and roof are added. This hive is now positioned to one side of the original site with the entrance facing at right angles to its original orientation.

5. After two days, the parent stock is moved to the opposite side of the hive containing the artificial swarm, with the entrance facing in the opposite direction.
6. After a further two days, the parent stock is placed in another part of the apiary or removed to a new site.

Demaree System

Robin Tomlinson and the late Terry Pearson of Leeds were keen exponents of the Demaree system of management. However, like a good number of enterprising beekeepers, they preferred to work to what is known in Yorkshire as the 'reversed' Demaree system. It is also known as the 'modified' Demaree. To this end, it would be useful to explain the differences in purpose and function between the original method of swarm prevention and honey production as devised by GW Demaree, published in the April 1892 edition of the *American Bee Journal*, and the variants of the original plan.

Over the years, the Demaree system and its variants have been used by commercial and amateur beekeepers alike. These methods are, in some quarters, as popular now as when they were first introduced all those years ago. The Demaree system can be used in any of the popular hive configurations. It is most probably true to say that commercial producers, with large apiaries and using large-sized frames and boxes, find it easier – in time, labour and equipment – to use the de-queen–re-queen method of swarm control rather than a management system like the Demaree.

Demaree's Original Method

The system devised by Demaree is basically very simple. It can be applied most readily to a colony occupying a single brood chamber where the bees are covering eight frames. A comb from the existing brood chamber containing eggs, unsealed brood and the queen, but no queen cells, is placed into a second brood chamber and frames of drawn comb or frames fitted with foundation are used to fill up the box. An empty comb is used to complete the complement of frames in the original brood chamber, or the remaining frames may be pushed together and a dummy board inserted at one end. The hive is reassembled so that the new brood chamber containing the queen is at the bottom of the hive. A queen excluder is placed above this, followed by the supers, a second queen excluder and then the original brood chamber, inner cover and roof. A second queen excluder is required to keep drones out of the supers. Provision is made for the escape of drones that will inevitably become trapped in this upper chamber.

The original plan has been adapted with many variations over the years but remains basically the same.

Drawbacks

The original method has two drawbacks. It is not completely reliable once swarm cells have been started, particularly if these have become advanced. Also, it is not satisfactory if a heavy nectar flow starts before the enlarged hive is adequately filled with bees. These points must be heeded for success.

We know, through the early research of the late Dr Colin Butler and his successors, that pheromones are emitted by

The original Demaree method of swarm control

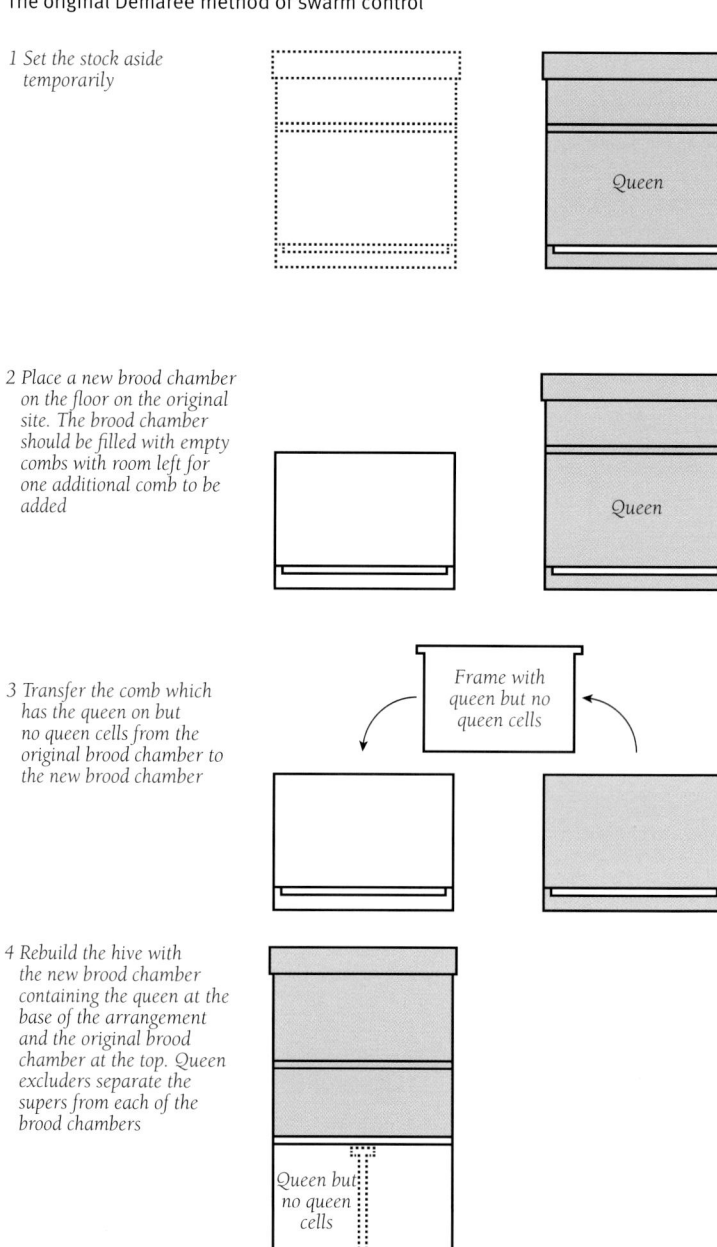

1 Set the stock aside temporarily

2 Place a new brood chamber on the floor on the original site. The brood chamber should be filled with empty combs with room left for one additional comb to be added

3 Transfer the comb which has the queen on but no queen cells from the original brood chamber to the new brood chamber

Frame with queen but no queen cells

4 Rebuild the hive with the new brood chamber containing the queen at the base of the arrangement and the original brood chamber at the top. Queen excluders separate the supers from each of the brood chambers

Queen but no queen cells

Bell Heather Honey Production

Demaree and reverse Demaree arrangements

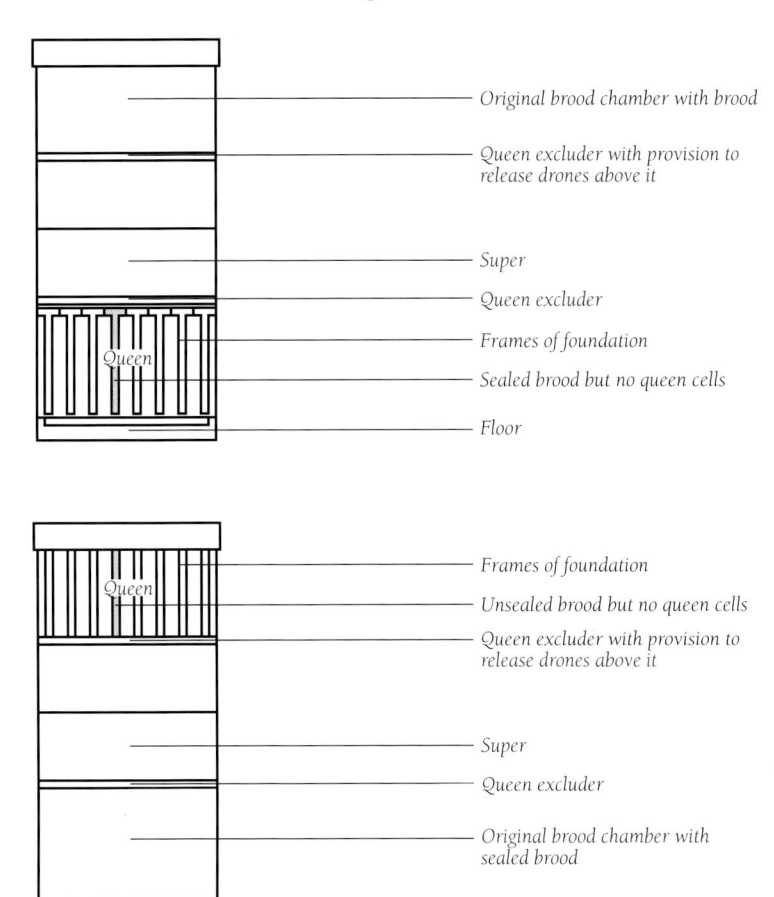

- Original brood chamber with brood
- Queen excluder with provision to release drones above it
- Super
- Queen excluder
- Frames of foundation
- Sealed brood but no queen cells
- Floor

- Frames of foundation
- Unsealed brood but no queen cells
- Queen excluder with provision to release drones above it
- Super
- Queen excluder
- Original brood chamber with sealed brood
- Floor

the brood that are known to influence the mechanisms of swarming. It should be understood that the Demaree method provides a check to brood production for a time and this will be quite marked if foundation is used rather than drawn comb. Although this is not always a disadvantage, there are occasions when a check in brood rearing would be undesirable, for example in the early part of the season. Division of the brood according to the Demaree system allows the bees and queen to stay together as one unit, more or less, and there is no guarantee that the colony will remain in the hive. The risk of swarming, in spite of the rearrangement of the hive, is indeed very great, especially if there are queen cells with queens near emergence and if scout bees have located a new nest site. Breaking down the queen cells will not always prevent a swarm leaving. Some would say that it is advisable to clip one pair of the queen's wings if the conditions described prevail.

It would be safe to think the following about the simple Demaree system:

- it is not a satisfactory method of swarm control for hives larger than the Modified National here in the British Isles
- it should be regarded as a method of swarm prevention to be applied before swarm cells are produced
- it should not be used at the beginning of, or during, a heavy nectar flow unless the enlarged hive is still brimming with bees
- there is a strong tendency for brood production to be checked with a subsequent shortage of foraging bees

- if foundation is used to fill up the brood chambers, care should be exercised to see that there is no shortage of nectar which would delay the drawing of foundation; feeding may be necessary.

Most beekeepers have no desire to rearrange a colony drastically unless there is a need to do so. There is a great deal to be said for the age old adage that 'there is no point in meeting trouble halfway' and little sense in dislocating a strong and satisfactory working unit just because it might swarm; it might not. Those who hold this view will presumably allow a colony to develop normally until they find swarm cells present. Then they will act.

The Reverse Demaree

The reverse Demaree addresses the disadvantages of the Demaree system. It is better to rearrange the colony so that there is a complete reversal of Demaree's original suggestions.

Two brood chambers are again required. Only sealed brood goes into one of the chambers which will go at the bottom of the rearranged hive. It is necessary to search this box for queen cells after about seven days, weather permitting or not. The queen remains with the unsealed brood and this is placed at the top of the hive, over a queen excluder. The honey supers may go between the two brood chambers, in which case a second excluder is inserted. Foundation must be used to complete the top brood chamber. Drones must be allowed to escape from this top chamber at intervals or through a second entrance to the hive. This arrangement should be maintained until the swarming impulse has ceased, the colony settles down and it stops producing queen cells, at which point the brood boxes can be interchanged. It is, of course, desirable to destroy any queen cells as the hive is rearranged. The brood should be inspected after seven days and any subsequent queen cells that might have been produced should be destroyed.

Checking for Queen Cells

When checking brood combs, it should not be taken for granted that there are not any newly laid eggs within slabs of sealed brood. A vigorous queen will find cells on the periphery of brood combs in which she can deposit eggs; this is a situation where the 'best laid plans of mice and men' go wrong. In every situation which involves destruction of queen cells, you must be diligent. This means, of course, that if you find eggs on a comb and you move this comb well away from the queen's laying area, queen cells may be produced as a division in the brood nest may have been created. Such queen cells must be destroyed well before eight days from when the egg was laid or plans will go wrong.

So many beekeepers have come to me saying that, having carried out swarm prevention methods literally 'to the book', the exercise has failed. It is in situations like the one described above that plans fail. The text books make no mention of such observable facts, so forewarned is forearmed.

Demaree's Intention

The purpose of the Demaree system is to prevent congestion in the brood chamber, thus keeping the queen actively laying.

My own experiences of the system suggests that it needs to be implemented during early to mid May (always dependent on colony strength, weather and the season). If undertaken later in the season, the honey crop is seriously diminished.

The most important point that must not be overlooked before a colony is 'Demareed' is the condition of the queen. She must be laying freely and in full vigour. A failing queen will give problems in a relatively short time. I have also found that the further away the queenless brood is placed from the queenright brood box, the greater the check on the colony's nectar-gathering powers, no doubt as a consequence of brood pheromone distribution within the hive. When the Demaree method is used, queen cells are usually built in the queenless box of brood. These queen cells should be destroyed no later than eight days after the colony has been Demareed. In most cases, these queen cells are inferior; the queens produced are at best 'scrub', although they could be considered of use if nothing else is available and you have a need to make up nuclei.

Note: The reverse Demaree's success comes about because the use of foundation checks the queen's egg laying as there are no suitable brood cells immediately available in which to lay. The swarming impulse abates because of the absence of pheromones which stimulate the production of queen cells.

THE YORK METHOD FOR DEALING WITH LATE-SEASON SWARM PREPARATIONS

In 1979, I was fortunate to visit the queen rearing operations of Harvey York Jr, York Bee Farms, USA, spending a week with him to see his operations. His aim was to get his colonies to swarm as early in the season as nature would allow them to.

As is always the case in nature, colonies often build queen cells later in the season, especially if a colony has been artificially swarmed and the old queen has been kept, usually to retain her inherent qualities. With the build up of a colony, the queen's deficiency in providing sufficient queen substance triggers the swarming impulse. Swarms late in the season are anything but desirable.

At York Bee Farms, I was able to witness a most novel process first hand. Upon finding queen cells in a brood chamber, all were ruthlessly destroyed. The brood chamber was reorganised with a frame of stores, followed by alternating frames of brood and foundation to fill the box. This arrangement has the effect (as was explained to me) of confusing the bees because of the divided brood nest. The bees, in their confused state, settled down and drew out the foundation and the queen laid eggs in these newly prepared cells. At some stage, supersedure cells were produced and, in one colony, I was shown two queens (mother and daughter laying eggs side by side). Eventually, the old queen was neglected by the nurse bees as the younger queen was producing more queen substance.

My colleague Charles Austin, in Barnsley, South Yorkshire, uses a similar method which he refers to as 'chequerboard' management. As I have explained elsewhere, putting a single frame of foundation in the heart of the brood nest, thus splitting it, can induce swarming preparations. Trying such a method could well be very chancy in the British Isles due to its fickle climate, although it appeared to work well in Florida.

The approach to swarm prevention and control may vary from district to district; it is best to try out such procedures with a fairly open mind. Any system needs to be refined by the operator over a number of seasons, so that its success can be carefully assessed in differing conditions. It is so easy to discard a system as no good without giving it due consideration. In any case, 'give it a go' as they say in Yorkshire, you may well find the system works for you.

PREPARATION OF COLONIES FOR BELL HEATHER ON THE MOOR

Colonies for bell and ling heather production are prepared for the moor in the same way. The brood combs within the brood chamber are arranged to follow the same pattern. See the following section on ling heather honey production and how to prepare colonies for the moor.

SYSTEMS OF MANAGEMENT FOR LING HEATHER HONEY PRODUCTION

The most critical factors for successful production of a surplus of ling heather honey are a large area of heather forage, good weather and a maximised number of foraging bees. Success demands all of the above factors and will not be achieved one without the others, and a bit of good luck thrown in. The duration of nectar flow from ling heather is, at best, 10–12 days, owing to changeable weather conditions, despite the bees being on the stances for three weeks or so. There are exceptions, particularly if there is prolonged good weather.

The bees are taken to the moors around 6–8 August and the main flow will generally be over by 28 August. This can be regarded as the norm, unless weather conditions earlier in the year delay the beginning of flowering time. Local variation can be expected. Beekeepers should not be lulled into thinking that continued good weather keeps the ling heather flowering; it does not. The bees will seek out other nectar-bearing plants such as rosebay willowherb. It is better to remove the crop of ling heather honey and then continue with supers of empty drawn comb for the bees to fill with, at best, a heather blend, than to risk the purity of the entire ling heather honey crop.

Experienced heathermen will tell you that, generally, the nectar flow starts on or around 8 August. Of the total flowering period, the weather typically prevents the flowers from yielding nectar for 10–12 days, resulting in a similar number of days of actual nectar flow. It is incumbent upon the beekeeper to maximise the available time that bees are able to forage. The siting of stances can and does influence how much time the bees can fly. The importance of siting should never be underestimated.

Need for an Adequate Foraging Work Force

It is mentioned elsewhere in the text and well worth repeating that to garner a worthwhile surplus of heather honey, it is an absolute necessity to have the maximum number of 'field' bees in the colony, those of the correct age to forage for nectar. The amount of nectar that can be collected is in direct ratio to the number of foragers the beekeeper can muster in his colonies prior to onset of the nectar flow. So many beekeepers overlook this vital point; most are more than satisfied with stocks if they contain several combs of sealed brood, without taking into consideration the size of the foraging force. The use of a management technique using nuclei comes into its own for late honey crops, especially working for the ling.

To emphasise the point more thoroughly, we need to consider the development of the bee from the laying of the egg until it becomes a field worker. This takes about 31 days.

- A worker egg laid by the queen takes three days to hatch.
- The larva is sealed within its cell after a further six days.
- The larva emerges from its cell as an adult 12 days later.
- The adult bee, following duties within the colony, becomes a forager after a further nine to ten days.

After this time, the foraging bee has become fully developed, ready to play her part as a true 'honey gatherer'. Considering the calendar for the ling heather honey crop, the aim is for the beekeeper to have eight to ten combs of sealed brood by the end of the first week in July. Brood sealed around 8 July will emerge around 20 July with a further ten days for the adults to mature and become foraging bees ready for field duties in the first week of August. Eight combs of brood would yield some 30,000 foraging bees while those previously reared would approximate to 40,000–50,000 bees. Following the summer solstice, the days begin to shorten by one minute per day. An established colony responds to reducing day length by contracting in size, whereas nuclei are driven to continue to expand as a natural survival strategy.

Under similar management conditions, bees worked for earlier crops in early June, given favourable weather and no swarming problems, could well yield good crops of lime, clover, blackberry, borage, or rosebay willowherb.

Flowering times favour the gathering of such crops as the days are still lengthening. Colonies are readily able to build up, as is the natural life cycle of the colony. Earlier crops of oilseed rape, sycamore, horse chestnut and possibly hawthorn, coupled with pleasant weather, all give succour to increase the queen's egg-laying rate resulting in peak colony strength and honey-gathering capacity when the crops are in bloom.

Under normal conditions, the queen's egg-laying rate begins to reduce in late July with the shortening day length and the reduction of available nectar sources. The colony will give every indication that it is teeming with bees. Three-quarters of them are worn out and they would be finished

Ling Heather Honey Production

within a week on the upland moors. A bee's life depends on the work it has performed; like all creatures it, too, has a lifespan geared to effort. Therefore, colonies destined for the heather require a different management technique in order to prosper on the moors. It is understandable that colonies which have given their all through the summer are a spent force. This is one of the primary reasons for poor returns from the moors if special forms of managment are not adopted.

PREPARATION

There are a number of ways of getting colonies ready for ling heather honey production. The secret is not to forfeit the summer crop, so easily possible if using ill-thought-out schemes. So beware. First-timers need to understand that ling heather honey production requires a totally different technique from that used for summer floral honey. Working for ling heather involves more time, more equipment, more colonies of bees and, above all, more effort. The reward is a fine honey that makes the effort worthwhile. However, despite all the work and effort, actions may well come to nought. The heatherman is rarely despondent as he or she knows that the bees seldom return from the moors without their winter's keep. Heather stores are a natural food, abundant in pollen, and excellent as such to support early brood rearing the following spring.

There are numerous ways of getting stocks ready for ling heather honey production. However, experienced heathermen have their own systems which they will use to good effect. There are three favoured approaches. These are:

- the 'forced' swarm method to produce a nucleus
- the use of swarms
- the nucleus method.

In addition, these two systems are worth a try:

- the use of the Horsley board
- a system for working late crops such as clover, blackberry, borage or ling heather.

For migratory heather honey production, the use of standard Modified National hives is recommended, generally restricting brood area to a maximum of ten brood frames.

THE 'FORCED' SWARM METHOD

The 'forced' swarm method relies on the beekeeper forcing the colony to raise queen cells by restricting the space available for its expansion in the early part of the season. This is a common situation that confronts feral colonies living in confined spaces. For the purposes of the practical beekeeper, the system relies on the bees being restricted to a single brood box. Although the use of double brood chambers will suffice and stocks always overwinter better in this arrangement, the time it takes for queen cells to be drawn will be that much longer.

Colonies wintered on two brood chambers should be confined to a single brood box some time from early to mid February (season dependent). By early spring, the bees will generally be occupying the top box and any brood rearing taking place will be in this chamber. On a suitable day, the beekeeper can carefully remove the mouseguard, take the bottom box from the floor board with minimum impact upon the colony and place the top box in its place, always providing there is no brood in the bottom box. If brood is found in the lower box, the arrangement is left intact. Should the bottom box be free of brood, all will be well to proceed. Place this box, complete with its bees, onto the crownboard with another quilt or crownboard with the feed holes covered above. The remaining bees will move down during the cold of the night, allowing the beekeeper to remove this box completely the following day and replace it with an empty super.

Should the colony be found to be short of food, it can be fed with fondant or 'thin' syrup (500 g sugar:625 ml water) in a contact feeder. A simple contact feeder can be made from a discarded cleaned catering-size coffee container. This is ideal as a throw-away feeder. Simply punch a series of very small holes, 20 or more, in a 30 mm diameter circle in the centre of the lid. Fill the container with syrup and press home the lid tightly to ensure a vacuum can be created. Invert the container over a spare container until it stops dripping and then place it over the feed hole in the crownboard (making sure you do not spill syrup). Rapid feeders are of no use in spring as the bees will not leave the cluster to collect the liquid, whereas with a contact feeder, as the name implies, the bees will take liquid that is in contact with them with alacrity. Once feeding commences, brood rearing will start in earnest and feeding must be continued until the bees are able to forage actively for natural food sometime in early April.

Very soon the colony will commence drone production. Coinciding with the lack of brood space, this will induce swarming. Once these drones are on the wing, queen cell production will start. Close observation by the beekeeper is essential. The presence of queen cells gives the beekeeper the opportunity to make up a five-frame nucleus from this stock, using one of the available queen cells. The method of making up a nucleus is discussed elsewhere.

THE USE OF SWARMS

For a number of years, I collected swarms and casts, building up such stocks ready for uniting. Always provided they were disease free, these stocks were united to make up strong stocks just prior to the heather coming into flower. An area in the apiary was set aside for this purpose.

Collected swarms are kept cool until late in the evening, when a large white sheet is placed in front of a prepared hive consisting of a floor board, a shallow box (super) containing frames of drawn comb, a deep box with two frames of drawn comb placed in the centre and frames of foundation to either side, and a crownboard with its feed holes covered. The object of the shallow box is to ensure that the frames of foundation in the deep box above are properly drawn to the bottom bars of the frames. The two drawn combs in the deep chamber instigate the drawing of the foundation by the bees and the queen will be attracted into this chamber by the newly drawn comb.

Two or three swarms are literally dumped en masse together on the sheet to unite them. Look out for all the queens and select the best (youngest) of them. A simple method of gauging the age of a queen is by looking at her wings. If these are frayed or rough in any way, this is a good indication of an ageing queen. Looking for the queens as they make their way into the hive needs absolute concentration; it is so easy

Dumping two swarms onto a sheet in front of a hive and looking for the queens in order to select the best one [Charles Bell]

to be distracted by the sight of the bees fanning out. The bees knowing instinctively to march into their new headquarters. It is simple to miss a 'slimmed down' queen amongst a mass of bees, even if she is marked. She is very often seen climbing over her charges as she heads to the hive entrance to get away from the light.

No matter what, these stocks should only be sent to the moor after uniting them to an existing colony and rearranging the brood (described elsewhere in the text), otherwise they will lack a strong foraging force. Any surplus brood that becomes available at the time of this rearrangement can be given to other stocks or nuclei. Surplus brood is always a blessing.

Making Use of Surplus Queen Cells from a Swarmed Colony

The easiest method available for obtaining a ripe queen cell is to obtain one from a colony that has recently swarmed or is about to swarm. The chosen queen cell should be taken from that part of the brood frame furthest away from any drone brood. This is because the nurse bees that feed drone brood do not produce the same quality of royal jelly as is required for good queen larvae. It is extra insurance to ensure that the queen in the chosen cell has been well fed with the most nutritious royal jelly The queen cell chosen should be one that has the appearance of a peanut. A queen cell with an exterior that looks as though it has been shaved down is ready for the queen to emerge and this is a sure sign that there is a queen present. The cell should be removed carefully, complete with a 'heel' of wax to enable it to be affixed into its new position. It should be attached to the centre of a frame of sealed brood where the bees would expect to find a queen cell in natural conditions. This frame should be situated in the centre of a made-up nucleus where the attendant bees will nurture the cell in readiness for the virgin queen to emerge.

THE NUCLEUS METHOD

By definition (under the now defunct British Standard Specification 1372:1947), a nucleus colony is a colony of bees occupying not more than five British Standard frames. It is, in fact, a miniature colony and should contain a balanced proportion of bees (including drones), brood and food. As a commercial unit it should be headed by a current season's queen.

In late May or early June, make up a nucleus from each strong colony, or one or more colonies may be divided to get the same result. I consider my preferred method for making up a nucleus is the best for the average beekeeper.

- The nucleus should contain two frames of sealed brood and one empty drawn comb.
- There should be sufficient bees to cover a minimum of four combs.
- A ripe queen cell should be added to the nucleus by securing it between facing surfaces of two combs of sealed brood, a natural location for a queen cell. Positioned here, the nurse bees will be nurturing the surrounding brood and the warmth will help the queen cell to mature.
- The nucleus hive entrance should be kept closed down to a minimum.
- A small inverted contact feeder of thin sugar syrup (500 g sugar:625 ml water) should be placed over the crownboard to assist the bees. This feeding should be carried out when the bees are not flying. Avoid spillage at all costs.
- The nucleus should be left well alone, other than watching for bees entering the hive carrying pollen.
- After two weeks or so (weather dependent), the young queen should be in lay. This will be apparent when there is a large quantity of pollen being brought into the hive and by the highly active manner of the incoming pollen foragers.

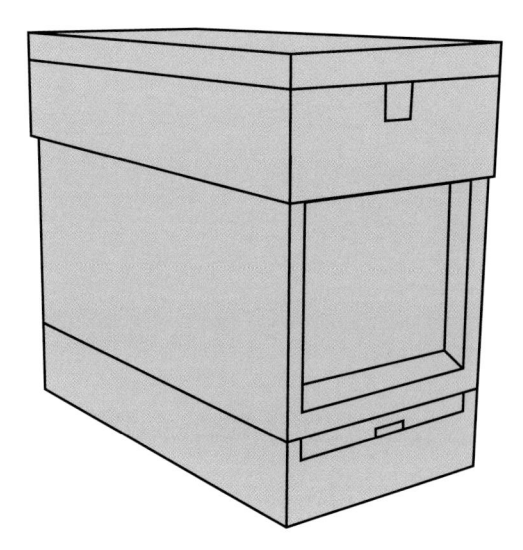

A nucleus hive accommodates fewer frames but is the same as a full size hive in all other respects. Entrances should be reduced to a single bee space to minimise robbing

A dummy board, sized to fit the hive, can be used to contain brood frames, for example in the making up of nuclei in a full-size brood box

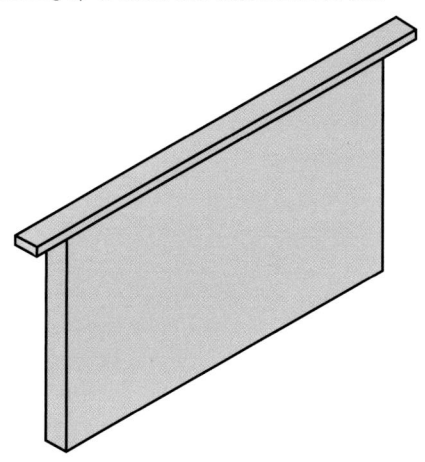

- Should the weather turn inclement, the feeder should be topped up at dusk when bees have ceased flying.

To enable the nucleus to expand to a full-size colony, with the queen in lay, it should be transferred into a full-size brood chamber with suitable packing to reduce the likelihood of chalkbrood and chilled brood. Fill the empty area with sacking/packing. As the colony builds up and develops, additional drawn combs should be added as required, resulting in a good colony of around ten combs by the beginning of August.

Do not be tempted to take such a nucleus colony as it stands to the moor. It should be united to an established stock to get the maximum results from a large foraging field force. This is most important. At the making-up stage, the nucleus should be placed in close proximity to an established strong colony in readiness for uniting them together. Once united, the brood should be rearranged as detailed later in the text.

General Procedure for Making Nuclei for Increase

To make up nuclei for increase is a relatively simple procedure. They should be made early in the season. It is essential that some simple principles are adopted. Increase must only be undertaken when drones are on the wing; colonies to be divided must have their own drones.

It is best to make increase when bees would naturally be in a ready state. I have found that it is best to make divisions earlier in the season, rather than later, as I have found that aged drones are not as fertile as younger drones, which are always preferred.

The making up of nuclei requires a certain amount of discipline and patience in that, once created, they should be left alone. No good will come of looking frequently to see if the queen is mated. Simple observation at the hive entrance can give a good indication of how the nucleus is doing by the amount and size of the pollen loads being taken in and especially the vigour and the rushing back and forth of the foragers entering the hive. A good yardstick of when to look into the brood nest is about ten days from making up the nucleus, but if the weather has been poor, this inspection should be delayed.

Method 1

The easiest and simplest method to make up a nucleus is to use a colony that is about to swarm, or has recently swarmed but has not yet issued any casts. For purposes of both increase and for use at the heather, a colony that is about to swarm is ideal.

- You will need five-frame nucleus boxes or brood boxes closed off with dummy boards.
- Divide the swarmed colony into the units required by having two combs of sealed brood plus two combs of food stores, ensuring that a queen cell is present in each. Add one empty brood comb or a frame with foundation to each nucleus.
- If using a brood box, empty space should be carefully packed with sacking to keep the nucleus colony warm.

- The combs within a nucleus should be well covered with bees. When made up, remove the nucleus to a new stance in the apiary, always in shade.
- Entrances should be restricted to a single bee space. Roofs should be bee-tight for protection from robbing bees. The bees should be confined to their new quarters for three days; plug the entrances tightly with grass.
- After three days, grass plugs should be loosened to help the bees to release themselves. This should be done while it is dark. When the bees are allowed to leave the hive after a period of confinement, they learn their new hive position and entrance location.

Method 2

Another method for making up a nucleus makes use of a virgin queen.

- Make up a colony of three combs, one of food, one of sealed brood and one empty comb or a frame of foundation.
- Shake three combs of bees into the made-up nucleus, confining the bees immediately.
- When it is dark, shake the nucleus, sufficient to dislodge the bees from the combs, and then drop the virgin queen into the mêlée. In the confusion, the virgin queen will adopt the hive odour very quickly with the bees settling down to find they have a queen amongst them. The bees should subsequently

be kept confined for at least three days by plugging the entrance of the hive with grass, as above.
- After three days of confinement, release the grass plug sufficiently that the bees can find their way out, again when it is dark. A small feeder should be in place, in case the weather turns inclement, but always feed the bees in the evening.

Method 3

A further method is to make up a three-comb nucleus in the middle of the day (when the bees are flying).

- Take two combs of sealed brood with bees and a comb of sealed stores from an established colony, taking care that the queen is not on any of the combs.
- Shake additional young bees into the nucleus. Make sure you do not transfer the queen.
- Add a ripe queen cell obtained from another colony. The cell should be placed as described previously.
- Confine the bees to their hive for three days. Once released, restrict the hive entrance to one bee space.

Dealing with a Swarmed Colony with Emerging Queens

In my formative years, I was reminded year after year that it is not a sin to have your bees swarm, but a real sin if they are allowed to cast because the swarmed colony has not been dealt with. Very often, a beekeeper will find that a colony has swarmed, new queens are emerging and he or she is at a loss as to what to do.

If increase is not required, the queen cells can be fixed into queen cages for use by colleagues. If that is not a suitable alternative, the following is the simplest method to adopt.

- Inspect the frames until you find those brood areas with emerging queen cells.
- Carefully open the queen cells to release the imprisoned queens (the queen cells will have been shaved down by the nurse bees to assist release; this shaving down is quite obvious).
- Let the queens run onto the comb.
- When all the queens on a comb have been released, shake all the bees off it so that they drop into the hive, creating a mêlée.
- It is suggested that you give the bees plenty of rough treatment through shaking; the general disruption arrests any intentions to cast because of the wholesale confusion.
- As the bees settle down, the released queens instinctively look for the others, with the strongest seeing off those that are weaker.

Within a day the colony settles down. Once the new queen is mated and begins to lay, the colony moves forward.

A SYSTEM FOR WORKING LATE CROPS SUCH AS CLOVER, BLACKBERRY, BORAGE OR LING HEATHER

This is a system suitable for forage that becomes available late in the summer season. It is a method of relative simplicity but may appear a complex operation. The principle is as follows:

- colonies are worked in pairs. In each case, colony A is strong in the spring and will be the non-heather colony. The second colony, B, is the weaker colony in the spring and will become the heather colony
- colony B need not already exist and may be made up at the right time as a nucleus from colony A as detailed subsequently
- unless oilseed rape provides a source of food, brood rearing in colony A may need to be stimulated with 'thin' sugar syrup (500 g sugar:625 ml water) and by 'doubling' this colony (see elsewhere in the text)
- the foraging bees of both colonies will be united for gathering clover, blackberry or borage nectar. The brood of both colonies will be united at the right time to give a population of foragers of optimum age for the heather.

Times and Rates of Brood Rearing

I have mentioned elsewhere the necessity to understand that, for the collection of any honey crop, only bees of foraging age during the actual period of the honey flow are of any real use. The other bees, as consumers, have a negative impact on the

Ling Heather Honey Production

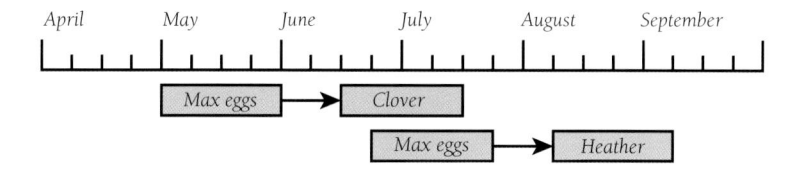

```
April     May      June     July    August   September
|....|....|....|....|....|....|
         [Max eggs] → [Clover]
                  [Max eggs] → [Heather]
```

The system for working late crops in schematic form

potential honey surplus. Therefore, it is essential to secure the maximum number of foragers during the time of each honey flow to obtain a surplus. I have found that the majority of beekeepers work hives that are too thinly populated with bees at the best times. Bees become foragers when they are at least two to three weeks old, with an average working life of up to six weeks. Therefore, the majority of bees should be two to three weeks old at the beginning of each nectar flow.

Taking clover as an example, flowering from around the third week in June. It will be seen that to take advantage of the flow, the maximum number of foraging workers should be emerging as adults at the beginning of June. Going three weeks further back (since workers take 21 days to develop), egg laying needs to be maximised around the middle of May and continue into early June. With ling heather, flowering the second week in August, egg laying needs to peak around the third week in June and onwards. It is impossible to secure satisfactory harvests from both crops from the same population of bees.

Method

Immediately after the first spring inspection, by careful movement of hives over short distances at a time, arrange the colonies in pairs, each pair of hives within 0.6 m (2 ft) of each other and at least 1.8 m (6 ft) between pairs.

- Colony A should be stronger and colony B weaker.
- Colony B can be weakened if necessary (if it progresses rapidly) and colony A strengthened by transferring combs of emerging brood (but no bees) from B to A.
- Colony B must be prevented from building up too rapidly and too early, which it will naturally try to do, especially if there is a nectar flow and the weather is good.
- Alternatively, colony B may, at this time, be just an empty hive with a nucleus added later.

During April, feed colony A if necessary (there is no need if there is a flow from oilseed rape). Use 'thin' sugar syrup (500 g sugar:625 ml water) fed via a contact feeder. If colony B has plenty of stores, do not feed it at all. During the last week in April, undertake doubling of colony A (unless it has been wintered as such) as described elsewhere in the text. This will require heavy feeding of 'thick' sugar syrup (1 kg sugar:625 ml water) in most situations to get fresh foundation drawn into comb. This operation will encourage the queen to increase her egg-laying rate. Do not continue to feed if nectar is coming in; if there are signs of a heavy nectar flow, add a queen excluder and a super over the two brood boxes.

In the situation where there is, as yet, no colony B, the time to extract a nucleus from A is when the bees are covering 15–16 combs, probably in the second to third week in May.

[A] [B]   [A] [B]   [A] [B]

The arrangement of hives when operating the system for late crops

Make up the nucleus as follows.

- Take out four combs from colony A, two of emerging brood and two of stores, ensuring all are well covered with bees and the queen is not amongst them.
- Transfer the combs to hive B, which may be a nucleus hive or a full-sized brood chamber contracted with two dummy boards and packing. Shake in bees from a fifth comb to make up for the shortfall of flying bees that go back to colony A.
- Close the entrance with a plug of grass but not too tightly. This ensures young bees mark the entrance when clearing out the grass plug. Add a small contact feeder containing 'thin' syrup.
- After four days, add a ripe queen cell, or introduce a laying queen after two days. Check to ensure no emergency queen cells have been made; if so, destroy them all. Introducing a laying queen is relatively safe and straightforward when a colony has been queenless for this period of time. A Butler cage is the ideal method. A word of caution: if it is a cage that has never had a queen caged within it, I recommend pouring boiling water over it before use, to remove

any unnatural odours. If time permits, place the cage over the feed hole of a crownboard in order that it may take on a hive odour. Any queen with a strange odour may be readily 'balled' when released. The queen should be placed in the cage and the end covered with a small piece of newspaper held in place with a rubber band. The caged queen should be placed in a position where nurse bees would expect to find her, that is between brood combs. The nurse bees will be attracted to the queen and will quickly begin to feed her through the cage mesh while others will be chewing through the paper to release her. To all intents and purposes, the queen will have adopted the same smell and exchange of pheromones will ensure unimpeded release and acceptance by the receiving colony.
- Rearrange the brood chamber in colony A, placing new frames of comb or foundation on the flanks of the brood nest to fill the gaps left by the removal of frames to make up the nucleus.
- The loss of brood will be made up quickly, making no difference to the honey gathering strength of the stock.

Whatever system is adopted, colonies A and B may now be left alone until at least the second week of June. The larger stock needs to watched for signs of swarm cell production. If none appear, all well and good, do nothing. However, if queen cells are started, the best solution would be to adopt a method of swarm control which maintains the colony as a single (split) unit, such as the use of a Horsley board or a Snelgrove board.

Ling Heather Honey Production

Using the System for Working for Surplus Honey in a Clover District

The next operation coincides invariably with the onset of the clover honey flow. If, in any district, there is no marked time when nectar becomes rapidly and suddenly available, the second week of June can be taken as the right time to proceed. At this time, when there is a good nectar flow, rearrange the brood in colony A, unless it has already been done for swarm control purposes.

- Put as much brood as possible in one of the brood boxes.
- The remainder of the brood (if any) stays with the queen in a brood chamber placed directly on the floor board. A queen excluder is placed between this box and the supers returned above.
- The box of brood is added above the supers. No type of screen board is used.
- The stock is left alone for three days to give it time to settle down.
- On the third day, the following changes are made:

    - Colony B is moved to the opposite side of colony A, moving colony A slightly sideways into the space that was between colony A and colony B.
    - The top brood box is removed from colony A and the bees shaken lightly but deliberately from each of the combs onto the super frames, below. The object of light shaking is to dislodge the older

flying bees; experience shows that young bees are more sure-footed and cling to the face of the comb.
- The combs of brood are replaced, along with any bees still clinging to these combs. Then, after opening the top of colony B, the box of brood from colony A is put on top of that of colony B.

By these operations:

- colony B now has the brood and the very young bees from both colonies A and B
- because of the change in positions of the hives of colonies A and B, all the flying bees will be working from colony A where they will forage for the clover surplus.

Points to note are:

- the yield will be considerably greater than that from two colonies each working on their own
- the colony engaged in honey production has very little brood to feed at the time of the honey flow and all the bees can get on with essential work
- neither colony is in a condition to swarm.

Proceeding to Harvest Ling Heather with this System

Colony B is now in perfect condition for bringing on for the heather harvest. It is early June; the stock has two boxes of

Ling Heather Honey Production

brood. As the bees in the colony reach foraging age, they bring in clover that encourages continued brood rearing. The queen is young and has had no chance to lay heavily as the colony has always been small. The change in conditions allows her to rise to the occasion now that there is incoming nectar and plenty of nurse bees. The adult bees at this time will be too old to work the heather, but by the first week in August the hive will be crowded with bees of just the right age as all the brood will have emerged and the resulting bees will be one to four weeks old. Thereafter, bees will emerge at the rate of 2,000–3,000 per day, most of which will have some opportunity to gather heather honey. At least three weeks before preparing to take the bees to the moors, a queen excluder needs to be inserted between the brood and supers. Once the combs in the top box are cleared of brood, the box can be removed and the bees shaken off into a super which is put in its place.

Simplified Timetable

The manipulations are stated as a general guide. Of course, there has to be some latitude in the dates, because of the weather, location and individual circumstances.

April 2–10	Arrange colonies in pairs, strong colony A next to weaker colony B, or strong colony A next to empty hive B.
April 2–25	Give colony A a stimulative feed or uncap sealed stores in the brood box. Do not feed colony B if it can be avoided.

April 25–30	'Double' colony A if not overwintered on two brood chambers. Continue to provide stimulative feed unless there is incoming nectar. Feeding may be essential, particularly if foundation is used. If colony B exceeds six combs of brood, give a comb of brood to colony A. Note: if new queens are required, make preparations for queen-raising in the strongest stock.
May 15–20	If there is no colony B, extract a nucleus from colony A. Introduce a laying queen on the second day or ripe queen cell on the fourth day. Super colony A as required. Watch for signs of swarming in colony A. Take measures if necessary to prevent swarming.
May 20–June 10	Continue to watch for signs of swarming in colony A. Encourage colony B to build up. Rearrange brood in colony A.
June 10–15	Three days later, change the positions of colony A and colony B. Take brood and very young bees from colony A and give them to colony B. Add supers to colony A as needed.

August 1–10   Condense colony B to a single brood box. Rearrange brood. Add supers, etc. Pack and send to the moors.

PROCEDURE FOR UNITING A NUCLEUS TO AN ESTABLISHED STOCK BEFORE MOVING IT TO THE HEATHER MOORS

An established colony should be strengthened with the addition of a nucleus colony by uniting the two together, ensuring there are ten combs of brood plus one of sealed food, and a large number of adult bees. It is most important that this is done, otherwise the foraging force will not be sufficient to gather the maximum surplus.

Provide a sealed comb of stores on the flank of the brood chamber. This should be sufficient to overcome any period of dearth should the bees be confined to the hive because of poor foraging weather.

The late Will Slinger always kept a number of brood combs containing sealed stores of heather honey for use on the moors. The combs were used on the flanks at the make-up stage, just before the move to the heather moors, as he was convinced that the smell of the heather got the bees in good heart and they would start foraging immediately, whereas those not prepared thus were that much slower to get going.

Finally, the brood must be rearranged at the time of uniting. This is really important if you want success. Frames of eggs and the youngest unsealed brood should be placed on the flanks of the brood chamber, with sealed and emerging brood placed in the centre of the brood nest.

Working from one side to the other, the combs should be placed as follows, assuming the use of an 11-frame brood box:

1. sealed stores (end comb)
2. eggs
3. eggs and larvae
4. larvae and sealed brood
5. sealed and emerging brood
6. sealed and emerging brood
7. sealed and emerging brood
8. larvae and sealed brood
9. eggs and larvae
10. eggs
11. sealed stores (end comb).

This arrangement will ensure that the queen continues her egg laying as the brood in the centre of the brood nest emerges. The unsealed brood towards the outside of the brood nest forces incoming nectar into the supers as there is no available storage space in the brood chamber.

The arrangment of brood combs for going to the heather

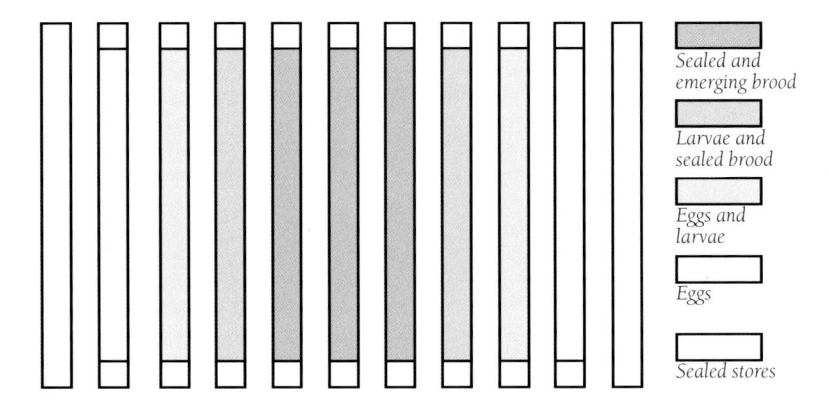

Sealed and emerging brood

Larvae and sealed brood

Eggs and larvae

Eggs

Sealed stores

Inserting a frame of sealed brood into the heart of the brood nest. The queen is able to lay in the comb once the bees in these cells have emerged [Brian Nellist]

Inserting a frame of eggs and young larvae at the flank of the brood nest forces the bees to store incoming nectar in the supers [Brian Nellist]

FINAL THOUGHTS ON UNITING

All uniting should be completed a few days before moving the colonies to the moor. Any sealed or unsealed combs of summer honey are removed. The unsealed combs can be given to other colonies with the sealed ones being processed for extraction. A super is added, which may contain drawn comb, foundation, starters, Ross Round sections or basswood sections. As an alternative to the super, a simple 'eke' or 'cog' may be used. The whole hive is suitably secured with straps, ready for shipping to the moor.

It is good practice to retain the queen from the established stock for use in making up a separate nucleus in readiness for uniting to the heather-gathering colony returning from the moor. Such colonies will benefit from reinforcement with new brood and bees. They will winter splendidly with the addition of young bees. Before uniting, the queen from the nucleus is removed (as she is the older of the two queens). A sheet of newspaper is placed directly on top of the brood chamber of the main colony and an empty brood chamber is added above. The combs and bees from the nucleus colony are transferred to this empty brood chamber with the combs placed towards the rear of the main hive, away from the entrance. The crownboard and roof are then replaced. The bees will chew their way through the sheet of newspaper over a period of a few days. There are other methods of uniting bees, but this is one of the safest methods for use by those new to such procedures.

Ling Heather Honey Production

USING A STOCK TO PRODUCE EXHIBITION HONEY SECTIONS OR CUT COMB

As a one-time exhibitor of section honey some 40 years ago, I adopted the following practice to ensure that the one colony I used had a very powerful foraging workforce. Early in the season, I placed the colony that was intended for the heather moor between two strong colonies (one at each side). After uniting a nucleus to the colony destined for the heather moors at least one flying day before departure, the two adjacent colonies were removed to another part of the apiary in order that their flying bees would join the colony bound for the heather. This colony was prepared with a section crate in place. The additional bees crammed into the hive forced the bees up into this crate. This colony was extremely strong; excellent ventilation with a full travelling screen was required for the journey to the heather. Because of the volume of bees, it was very often late in the evening before all of them were able to get into the hive, more especially if the evening was warm.

On two occasions, with hot weather, we had to envelop the bees in a loose-fitted cotton sheet, after carefully stopping up the hive entrance, as the bees would not enter the hive. The sheet was tied at the top. The hive and bees travelled quite happily on the back seat of my Rover car. Upon arrival at the moor, the sheet was carefully removed. All the bees travelled safely without ill effects. Return from the moor was straightforward. The toll on foraging bees was evident when the completed sections were removed prior to the return home.

TREATMENTS FOR VARROA

The issues relating to varroa treatment need to be addressed. There are several schools of thought on when is the best time to treat, whether before going to the moor or waiting until the colonies have returned. It is recommended that the beekeeper discusses his or her own situation directly with the bee inspector for their area.

MOVING BEES TO THE MOORS AND SITING HIVE STANCES

Most ardent heathermen make a trip to the moors a week to ten days before they move their bees to their moorland stances. I use the term 'stance' as it is a historical term used by heathermen in Scotland. Others may ask someone living locally to advise them of the first light flush of purple creeping over the hills.

My preference is that hives should be secured the night before moving in readiness for an unhindered start in the early morning. By doing so, if any snags or hold-ups are to be encountered, they can be dealt with there and then.

The method of securing hives can take many forms. These include the standard nylon hive strap, buckle and spiked cam hive strap and ratchet hive strap, all of which are wrapped

A spring fastener [Michael Badger]

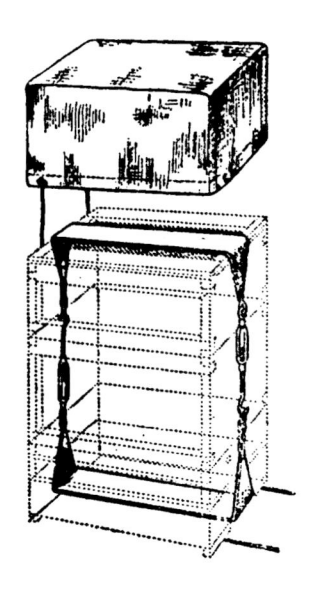

Old-fashioned hive clamp manufactured by Yorkshire Apiaries, Willerby, Hull, East Yorkshire, as featured in a 1946 catalogue

around the hive components and levered tight. Other methods of fastening include galvanised sheet-steel triangles, toggle fasteners, lock-slides, hive staples and spring fasteners, which are the simplest and most effective of all.

The entrance of the hive should be closed completely. A sponge foam strip of adquate depth is ideal for blocking the entrance; it is cut 3 mm longer than the width of the entrance to ensure a tight fit.

Care should be exercised to minimise the effects of travelling on the bees. Transporting hives using a well-sprung trailer or vehicle cushions the shock from poor road surfaces. The move will create some excitement that results in a rise in temperature within the hive. To minimise the danger of loss of bees as a consequence, a full-size ventilation screen should be set in place, replacing the crownboard, irrespective of the

Moving Bees to the Moors and Siting Hive Stances

distance to be travelled. Heat is then easily dissipated to the atmosphere. If daylight is visible, the bees are instinctively drawn to it which causes excitment and, ultimately, their death. Problems may arise when bees try to reach the light entering the hive through the screen; a mass of bees may reduce the air flow through the hive and the temperature may rise as a result. Using a mist sprayer, tepid water can be sprayed on the bees through the screen to help keep them from overheating. The provision of ample ventilation, space within the hive for the bees and travelling in the dark before dawn will see the bees arrive at the moor none the worse for the ordeal. Placing the hive within its upturned roof for travelling is common practice.

Old-type heather floor. Note the circular holes to the rear which were provided to hold phials of oil of wintergreen, a past remedy for acarine mite infestation [Michael Badger]

If you are taking a modest number of hives to the moor, they can be placed on the back seat of the car or in the car boot. Should it become possible for bees to escape from the hive, they will generally sit tight in the dark whereas, in daylight, they will invariably fly to the car window. They rarely cause much trouble unless there is a major incident, or the hive components come adrift or completely apart.

Commercial beekeepers may use specially constructed trailers; some are made with heavy damper suspension that minimises vibration. Modern trailers are generally well sprung, minimising vibration until off-road. If vibration could be an issue on the return journey from the moor, the crop is best removed beforehand to avoid the possibility of bees being drowned in liquid honey and the loss of many valuable bees.

When attached to the vehicle, the trailer bed should be horizontal, otherwise the hives will be pitched on an incline and the trailer may sway from side to side when being towed. This motion creates stress on the bees.

Before embarking on the trip to the moor, draw up a check list of essentials for all eventualities. The list should include protective gloves, boots and clothing, artificial bee-smoke aerosol, masking tape, a heavy-duty torch, spare hive straps, waterproofs and spare clothing. It is also essential that you travel with a companion, preferably a fellow beekeeper. Before departing, tell someone the location of your destination and the route you plan to take. Mishaps do happen and it is helpful to the rescue services to have an indication of your possible whereabouts. Use the Ordnance Survey grid mapping point to pinpoint your heather stance. Though not a necessity, a flask of tea/coffee and some sustenance will not go amiss, especially if you are going some distance. My

Purpose-made migratory beekeeping trailers. The design allows for the trailer wheels to be decoupled from the trailer platform to avoid potential theft
[Ivor Flatman]

colleague, Paul Snowden, advised me that on his first visit to the heather moors, he had become disorientated as to the exact location of the heather stance for which he was heading. So he advises taking an Ordnance Survey map with you.

Once you arrive at the moor, release the bees as soon as is practicable. It is essential that you get all hives ready and set level before opening the entrances. It is preferable to keep the bees imprisoned for about ten minutes once they have been set down at the stance in order to settle them. Once the hive is in place, carefully release the tension on the hive straps sufficiently to allow the travelling screen to be removed and replaced with a crownboard. If you are quick and deliberate, the removal of the ventilation screen can be done with minimum fuss. The bees are shaken off with a 'sharp thud' at the hive entrance, with little inconvenience to the bees or

Damage to dry stone walls caused by beekeepers, Rosedale, North Yorkshire [William Slinger]

the beekeeper. If you are producing cut comb or sections in particular, place a piece of good wool carpet the same size as the crownboard above it to retain the maximum warmth within the hive. Put on the hive roof and tighten the tension of the hive straps to secure the hive components together.

Do not be tempted to use heavy stones from nearby dry stone walls to weigh down hive roofs; it is disrespectful to the landowner and the stones are rarely replaced correctly.

At the time of returning from the moors, the whole moving procedure is repeated.

SITING HIVES

The siting of hives on their stances also needs consideration. Avoid communal stances, especially those where large numbers of stocks of bees may be set down, very often placed in straight lines and put closely together. This layout should be avoided, as bees are known to drift wildly when on the moors. The beekeeper should find an area for his or her sole use. This reduces the possibility of disease transmission from the colonies of another beekeeper. An individual heather stance should hold a maximum of 20 stocks. The hives should never be set out in a continuous straight line. They should be placed in pairs, 1.2 m from each other, with the next pair 3 m away. Hives are best placed on 25 x 25 mm battens, levelled up with pieces of stone or timber wedges, that keep the entrances close to the ground. This helps to keep the chill of the ground from the bees. The use of simple alighting boards keeps the vegetation at bay and helps heavily-laden bees enter the hive entrance rather than having to run down the front of the hive. In strong colonies, the entrance should be 250 mm wide x 10 mm deep, sufficiently limited to deter slugs and rodents from gaining access to the hive in poor weather. It is not unknown for adders to be found under hives that are placed directly on the ground. Therefore, caution is needed; wear stout boots and gloves when lifting hives for the return home.

Once the bees have been released, a final check should be made that all hive entrances have been unblocked to allow bees to fly. The late Bill Reynolds, a very experienced heatherman from Harrogate, North Yorkshire, recalled to the author how easy it is to overlook taking out the foam strip from a hive entrance. Bill was unflappable. Nevertheless, he was not shy in admitting how easy it is to be in error while unloading a large number of hives, or to be caught out with a dramatic change in the weather for the worse.

LITTLE FURTHER TO BE DONE WHILE THE BEES ARE AT THE MOOR

Thereafter, there is very little work to be done at the moor during the next three to four weeks. Should the weather be fine, new combs may be added after first removing full, capped combs, or partially sealed ones moved to the flanks of the super for completion and empty combs to the centre. *On no account should the brood nest be disturbed.* If the weather is exceptional, an additional super can be given above the first. It is better to give extra room at the top rather than below the existing super. Over the years, I have only had one or two occasions when three or more additional supers were needed for gathering pure ling heather honey. It has been stated previously that the average heather flow is about ten to 11 days, although there are always exceptions.

Hives in place at Bradfield Moor [Andrew Badger]

Moving Bees to the Moors and Siting Hive Stances

By the last week in August, the bees should be brought back to the home apiary without delay. Leaving the bees at the moor after the end of August is pure folly, as the heather flowers are generally past their best. This is when the first frosts appear that leave the flowers with a rusty look.

To ensure success at the heather, take only strong colonies. Insulation over the super is essential; nevertheless, the best packing is plenty of bees. Do not take colonies with more than ten combs of brood, or those on double brood or brood-and-a-half. Experienced heathermen will tell you eight to ten combs of brood will serve you well. You need a compact brood nest. The change of conditions stops brood rearing within a week. The object is to get surplus honey into supers, not to fill the brood area towards the end of the flow. The bees will begin to fill the brood area once all the brood has emerged. The bees' economic flying area can be up to five hectares, so it is essential to find a site that has 40 or more hectares of heather to work and where the bees are in the heart of the moor.

RETURN FROM THE MOOR

The return from the moors to the home apiary is best done quickly. It does not present the same difficulties as the outward trip for a number of reasons. The weather is generally cooler. The strong colonies will have weakened considerably through the loss of foragers. Nevertheless, the crop is best removed prior to moving the hives back to the home apiary to avoid the possibility of the bees being drowned in honey from those cells that have not been capped. The crop should be brought back to a warm room for ease of processing. It is essential that

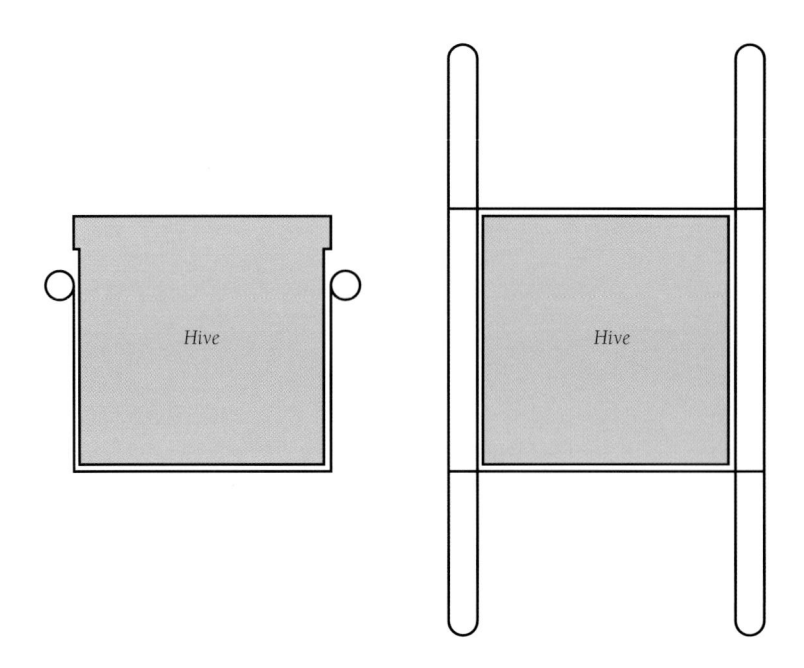

A simple carrying stretcher made up from two poles plus a heavy duty sack. When lifted, the stretcher clamps against the hive to stabilise it

the honey is not made too warm. I have found that excessively warmed heather honey has a disposition to 'foam' when it is pressed or extracted by centrifuge.

METHODS USED TO EXTRACT AND PROCESS HEATHER HONEY

There are various methods of extracting and processing heather honey. These can be summarised as:

- extraction by:
  - use of a tangential extractor following agitation of thixotropic ling heather honey in the combs
  - use of a centrifuge
  - pressing
- traditional square sections
- Ross Rounds or Cobana circular sections
- cut comb.

Beekeepers very often work with just one of these methods, while others venture into all of them. Those beekeepers making their first venture to the moors are advised to start off modestly by taking no more than two or three colonies. Very often, inexperienced beekeepers take more hives than they have properly prepared. One beekeeper that I know lost all four of his colonies the following spring as a result of not heeding the management rules for heather-going stocks.

THE EFFECTS OF SEASONAL VARIATIONS

The season, the weather, the colony, its prior management and the requirement to draw comb in which to store honey from foundation or starter strips will all impact on the potential amount of surplus. It has been stated previously that the weather is the main arbiter to success or failure. The 2013 season was affected by the weather conditions during the summer of 2012. That summer followed exceptionally warm weather in early spring that became extremely wet from the end of May until late July, with below average temperatures. The indifferent weather returned in mid August, with substantial rainfall throughout the autumn and into January 2013. The following March saw unprecedented heavy snow with north-easterly winds. Heavy frosts seriously damaged the heather buds. Growing conditions throughout April and into May were poor owing to low temperatures. Subsequently, the weather turned full circle and below-average rainfall brought drought conditions to the moors. The months of June and July ushered in extremely high temperatures that touched 30 °C in moorland districts. Yet the bell heather seemed to flourish despite these unusual conditions.

METHODS OPEN TO THE BEEKEEPER

Irrespective of terrain and access, it is best to remove comb honey from the hives before returning the colonies back to the home apiary. There is a major risk of damage to finished comb if it is left in place. Experience has shown that undulating moorland terrain, tracks and roadways are not suited to ordinary vehicles carrying heavy hives of bees with the honey crop still in place. Access tracks are full of potholes and ruts caused by farm machinery and off-road vehicles. Therefore, it is best to tranship the harvest by removing the supers or ekes from the hives to avoid damage to the combs of finished honey. Supers, ekes and section crates can be removed relatively easily by putting on clearer boards (eg, Canadian type) the previous day or, alternatively, through the use of artificial bee smoke in late afternoon. As the temperature is generally much cooler than at lower altitudes, the bees will not be flying at this time of day and those bees in the supers have generally moved down into the brood area. The traditional Porter bee escapes should not be used. They are liable to jam and can either prevent the bees from clearing or create a situation of two-way traffic. The latter results in the bees quickly moving any unsealed honey from the super to the brood chamber. My preference is for the Forrest-type clearer board as it has no working parts. Clearer boards must not be left on for more than 36 hours, otherwise the bees will find their way back into the supers and any unsealed honey will be removed in double-quick time. A note of caution. Should there be any brood in the supers, the nurse bees will not clear; their nursing instincts will cause them to stay with this brood.

Forrest clearer board [Michael Badger]

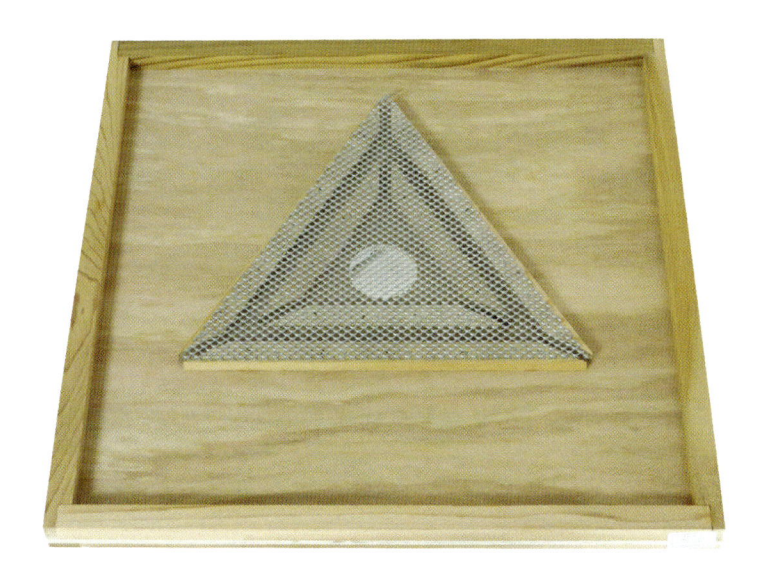

Having removed the super, I replace it with a shallow eke and ventilation screen to give the bees some room, as confining them to the brood chamber may create problems with a rise in temperature.

The reduction in overall hive weight by transhipping the honey crop makes it much easier to lift hives into the carrying vehicle. It can be difficult enough to manage large heavy hives as it is; those with supers in place will be extremely heavy.

During transit, large plastic horticultural trays are ideal for capturing honey dripping from damaged brace comb or unfinished combs. They are hygienic and can be washed easily. Any collected honey can be reclaimed for use at a later date for feeding to the bees or for making mead.

Extracted Honey

The majority of experienced heathermen work their colonies with a view to extracting the honey, seeing this as the best route to obtain the maximum surplus available to them.

The use of either narrow or wide spacing in supers is a matter of choice. The advantage of wide spacing is that the bees have fewer comb faces to cap. The bees will draw deeper cells to utilise the extra available space. Wide spacing can only be achieved with fully drawn combs because of the likelihood of burr comb construction in the gaps between frames when using foundation. The use of wide spacing is not always successful for ling heather production because of the fickle weather and its impact on the rate of incoming nectar. However, if the weather is good and nectar is coming into the hive in a large quantity, this might be an option if drawn combs are used. It is worth noting that a frame on narrow

Methods Used to Obtain and Process Heather Honey

Wide metal end

Narrow metal end

Groove for foundation in frame side bar

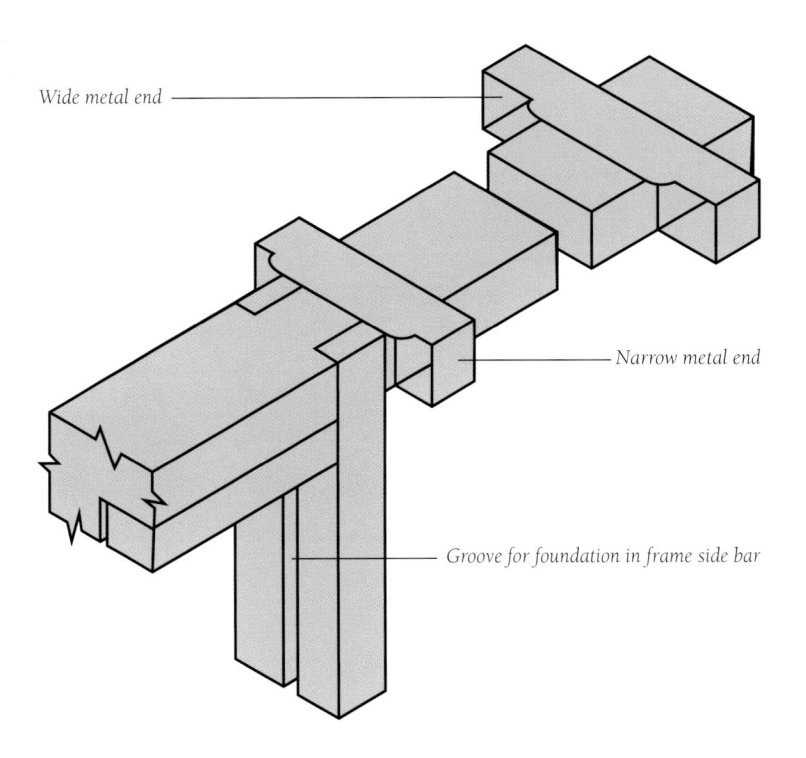

Narrow and wide metal ends

Narrow spacing is achieved by staggering wide metal ends (left) and is used when foundation is to be drawn, prior to spacing being increased (right)

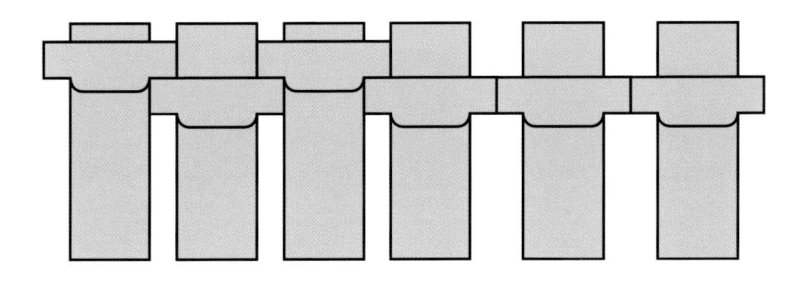

The use of half a sheet of thin foundation (cut diagonally) to produce a well-filled comb of heather honey. Note the drone cells to the lower right of the comb. If using this method, the frames of foundation should be placed with the diagonal cuts in alternating directions [Michael Badger]

A full crate of frames being removed from a hive. The wire across the top holds the frames of comb in place [David Pearce]

spacing is filled with approximately 1.25 kg of honey compared with just over 2 kg if it is on wide spacing. So, with floral honey production, it is well worth considering the use of frames on wide spacing.

Use of Drawn Combs

Those heatherman who are able to work early-season oilseed rape use this time to draw foundation for later use on the moors. It is a simple procedure that involves supering with frames of thin foundation, very often using 12 frames per super rather than the standard 11 frames (if using Modified National hives). An additional frame per super can be used equally in Langstroth or other hives. The reduced comb spacing ensures that the bees do not fill the gaps in between frames with brace comb. The beekeeper very often removes these combs once the bees have started drawing the cells, extracting what honey is available to ensure it does not granulate in the comb. These drawn combs are ideal for later use at the heather.

The use of partially-drawn comb over the 10–12 day heather nectar flow period ensures that the beekeeper gets the maximum return for his labours. It is an even better proposition if the beekeeper uses fully drawn comb that is free of any residual honey that may still be lodged in the cells, especially if oilseed rape was the previous crop.

Dealing with Partially Granulated Comb

Oilseed rape honey granulates rapidly, usually when the nectar flow ends in May. This, in turn, makes extraction difficult.

To remove a mass of granulated honey from combs, a simple procedure can be adopted.

- Take each of the frames of comb that need treating and fill any empty or partially-filled cells on the comb with clean water.
- Put these frames in another, empty, super.
- At dusk, put this super on the floor of the hive of a strong colony (one box only per colony), followed by the existing brood chamber and supers.
- Overnight, the treated colony will remove the granulated honey, using the water present within the combs to help liquefy the honey before it is stored elsewhere or consumed by the bees themselves.
- The bees find the super in this position unwelcome; it is unnatural for the colony to have wet combs of granulated stores near the the entrance. The bees remove the majority of the honey mass quickly so that within 36 hours it is completely cleared. There are exceptions. Should a small amount of granulated honey be left behind it can be carefully removed from the comb with a large dessert spoon and placed on a Pratley uncapping tray to liquefy it.
- The super can be removed and the hive reassembled. The cleaned up combs are stored in readiness for use later.

Use of a Wax Melter Cabinet

A fairly recent innovation is a heated cabinet that deals specifically with granulated honey. It can also be used for warming honey stored in bulk in plastic containers.

The unit is made from high quality stainless steel, supported on legs. Castors ensure complete mobility. Its dimensions are 990 x 490 x 340 mm (39 x 19 x 13 in) and it weighs 59 kg (132 lb). The tank is double-walled and fully insulated to a high specification. It has two separate thermostatically controlled heating elements that ensure honey and wax do not become overheated. The heating elements are strategically placed with one in the tank base and the other secured at the top, under the lid. The lid also incorporates a powerful fan that guarantees an even distribution of heat.

Dealing with granulated honey involves placing a quantity of filled combs, in their entirety, into the tank. Combs can first be cut from their frames if wished. No more than 120 kg (266 lb) of wax and honey should be melted in one cycle. The bottom heater is set at 40 °C with the top heater set at 80 °C. Simultaneously, the honey and wax melt under controlled conditions that will avoid damage to the honey. The different densities of the honey and the wax permit them to separate in layers. The honey sinks to the bottom of the tank, while the melted wax floats on top of it. The pure undamaged honey is drained off first, followed by the lighter yellow liquid wax.

Honey can also be recovered from cappings in a similar manner. Wax cappings can yield a significant quantity of additional honey when normal extraction methods are used. The unit is also ideal for bulk honey warming and for making both soft-set and seeded honey.

The capital outlay for such a device will certainly be recovered in a relatively short payback period. The manufacturers maintain that its correct use produces a honey yield nearing 100 per cent.

Heather Honey: A Comprehensive Guide

A wax melter in use [Michael Badger]

Using Specialist Extraction Equipment for Processing Heather Honey

The extraction of heather honey for small operators could, in the past, be achieved by the use of a Perforextractor, invented and made only by S Wolstencroft, Expert, BBKA, Bee Appliance Stores, Irlam, Salford, Manchester. The original item is no longer manufactured but was quite popular during the Second World War beekeeping boom. Alas, it fell by the wayside in the early 1950s. It is now an anachronism as most beekeepers use other methods. It was a simple device which consisted of a block of wood of suitable thickness, to which the frame to be agitated was clamped for stability, along with an agitator. The agitator was reminiscent of a wire brush in appearance, consisting of a block of wood approximately 300 mm (12 in) long x 50 mm (2 in) wide x 25 mm (1 in)

TRACTION OF HEATHER HONEY WITH PERFOREXTRACTOR

Honey from ling cannot be extracted in the usual manner, being gelatinous, but can be pressed from the combs by the use of a special press. This means often that drawn out combs, a valuable asset, are destroyed.

To overcome this difficulty we can supply a patent apparatus known as the "Perforextractor." This perforates the mid-rib of the comb, and the honey can then be extracted in an ordinary honey extractor, leaving the comb as clean and undamaged as if the honey had been clover honey. As drawn-out combs are usually valued at from 3/- to 4/- each, the cost of this apparatus is saved on the first few combs.

Complete outfit, with full instructions .... 45/- Post 1/- extra

The Perforextractor, Yorkshire Apiaries catalogue, 1945

deep with several banks of needles. The needle block was used to penetrate the cappings of the comb which are otherwise left intact. At the same time, the block is manoeuvred carefully back and forth, sufficient to agitate the thixotropic honey and change it temporarily to a fluid state. The comb was then ready for extraction using a tangential extractor. This method was tedious and only really suitable for a beekeeper with just one or two supers to extract.

The Use of The Norwegian 'Sjolis' Heather Honey Loosener

Serious operators opt for the Norwegian 'Sjolis' heather honey loosener. This is a very sophisticated piece of gear that is well worth the high capital outlay. It is hand operated and allows the operator to agitate the honey on both surfaces of a comb simultaneously. It automatically shifts the combs in either

Sjolis honey loosener [Willie Robson]

Industrial-scale tangential extractor [Willie Robson]

plane within the loosener's needle chamber to ensure all areas of the comb are treated. The agitators have small balls at the end of nylon pins. The operation ensures that a good stirring action takes place within the cells. The agitation process certainly speeds up extraction which is undertaken with a tangential extractor that needs to run at 450–500 rpm. The combs are turned several times to minimise comb breakage. Radial machines do not develop sufficient force for efficient extraction of thixotropic honeys.

One user of this equipment has designed a bee-proof cowl that allows the device to be placed on a colony of bees for cleaning purposes. The cowl avoids the potential problems of robbing, including the risk of bees robbing out diseased colonies in the frenzy that can be caused. For the purposes of the photograph the cowl has been removed to show the bees in action.

A METHOD OF CLEANING WET COMBS THAT SHOULD NEVER BE USED

A practice that should *never* be followed is to put supers of extracted wet combs into an open area of the apiary for the bees to clean up, no matter how tempting. Such practice creates great excitement; it most often gets the bees into a frenzy, especially when there is no nectar available at the end of the season when this unfitting practice usually takes place. The sudden appearance of honey in such conditions instigates robbing within the apiary. The bees become tetchy and sting all and sundry in the vicinity.

It is also a vector for disease. Once in this frenetic state, strong colonies will seek out weaker colonies that may have

Sjolis honey loosener being cleaned up by the bees with the isolation cowl normally in place for cleaning removed for ease of photography [Lester Quayle]

Methods Used to Obtain and Process Heather Honey

disease, or diseased colonies that have died out, thereby bringing back disease organisms from infected bees or combs into their own hives. Strong colonies are renowned for getting bee diseases through robbing activities, a view which the bee disease inspectors will support. In addition, such practices attract large numbers of wasps. At the end of the season, wasps are always on the lookout for easy pickings; wasps also demoralise weaker bee colonies if preventative measures are not in place. It is good practice to set up wasp traps made from large pickle-type jars. These should be placed some 20 m away from the apiary. The wasp traps are made by piercing the lid with a small hole of sufficient size for a wasp to enter. The vessel is half-filled with water and beer as an attractant. When the trap is full, it should be placed inside an airtight plastic bag and sealed so that the occupants are killed painlessly in readiness for disposal.

Bad practice – supers of wet comb left out for bees to clean up [Michael Badger]

A BETTER ALTERNATIVE FOR CLEANING WET COMBS

The cleaning up of wet combs should be done by placing wet supers onto beehives at dusk. The hive entrances should be reduced while cleaning up takes place as a precaution against robbing. The supers should be placed over a crownboard. A second crownboard is placed on top. Any gaps or holes in the hive woodwork are sealed up Overnight, the bees will clean these combs without getting overly excited. Generally, these combs will be ready for storage after 36 hours.

USE OF A CENTRIFUGE TO EXTRACT LING HEATHER HONEY

Some large-scale producers of ling heather honey extract their combs by scraping the unwired comb down to the midrib, either for pressing or for spinning out the honey with a high-speed centrifuge.

All beekeepers have limited time, so need to maximise efficiency. Cleaning up previously used frames or making up new ones is tedious, labour intensive and time consuming; such jobs are best reserved for 'wet days' and for the winter months. It can take at least two minutes to construct a frame, at best 30 frames to the hour. Any method to save on labour and effort is to be adopted readily.

The late Tom Bradford used a salvaged redundant ship's laundry spin dryer as a high-speed centrifuge, internally lined with stainless steel. It was acquired from a ship breaker's yard in Glasgow. The use of a centrifuge involves scraping the comb back to the midrib and putting the mixture of honey

Tom Bradford's heavy-duty centrifuge being used for extracting heather honey. The internal parts are stainless steel [Michael Badger]

and wax into a purpose-made cloth or nylon bag. The bag with its contents was placed in the drum of the centrifuge. This method introduces large amounts of air into the honey, with small air bubbles being uniform throughout. Care has to be taken that the honey is not too warm, or 'frothing' will occur which makes processing a messy business.

Once the bulk has been scraped down, the frames are placed in a radial extractor to clean them of the residual honey; it is amazing how much additional honey is obtained from such an operation.

A Smaller Scale Method Using a Domestic Spin Dryer

Some 30 years ago, I modified a domestic spin dryer for processing both heather honey and for recovering honey from wet cappings. The late Peter McScott of Harrogate obtained a number of House of Holland domestic spin dryers which came with stainless steel drums. These were modified for beekeepers' use to good effect. However, they will not work correctly without using a device that looks like a wooden corset-type stay and fits into the cylindrical drum. The 'stay' stops the linen scrim bag from blocking the holes in the drum. This is a very efficient piece of kit that has saved me hours of processing time.

Essentially, the honey and wax (avoiding stored pollen) is scraped from the midrib of the comb for processing and mashed up. Avoid spilling pieces onto the floor of the work area. The honey needs to be stored in a warm room prior to processing. The knack to working efficiently is to spin small quantities and not to overload the linen scrim bag.

The spin dryer is best elevated to a working height sufficient for a bulk container to fit underneath to catch the honey as it is discharged from the barrel. The bulk container should be allowed to fill to about 80 per cent capacity. This will allow entrapped air to rise to the surface; after 24 hours this can be skimmed off and the container topped up with honey similarly processed. It is essential that the linen scrim bag is strong and does not leak as the honey cannot be strained subsequently to remove bits of wax and other detritus. This bulk honey is then heated as described later to kill off the wild yeasts that are present in heather honey.

The process of centrifugal extraction introduces uniformly fine air bubbles that are very pleasing to the eye when the honey is packaged for sale in a jar.

Centrifuging can also be used to process wet cappings and, after spinning, leaves the cappings more or less bone dry and

The Author's spin dryer with the wooden 'stay' in place, which stops the linen scrim bag from blocking the holes in the drum [Caroline Timms]

ready for processing in a solar wax extractor, an MGA wax melter, or similar device.

PRESSED LING HEATHER HONEY

The beekeeper who decides to use this method will be sacrificing, in most cases, good quality drawn comb. The disadvantages of doing so have been explained elsewhere.

The pressing process involves cutting out the whole section of comb from within the frame in a large slab to avoid wastage. Combs of heather honey which have not been capped are processed separately from the capped heather honey and identified as such in bulk containers. This is because the honey which has not been capped has a higher water content which increases the risk of fermentation. The slabs of comb are placed in a strong cheesecloth or linen scrim positioned between

Peebles-type honey press (left), originally owned by Willie Smith of Smith hive fame, alongside other old types of honey press on display at Chain Bridge Honey Farm, Northumberland [Michael Badger]

the two faces of the heather press. The operator applies steadily racked up pressure to force out the honey through the cloth or scrim. The pressure involved will strain honey through the cloth free of wax and other detritus collected by the bees.

Despite cutting out the mass of comb with its contained honey, there is a small amount of honey that remains on each frame. The wet frames can be spun in a radial extractor to obtain the final dregs. To a commercial man this final process can yield quite an additional amount of honey. Any cappings are run through a centrifuge, leaving them clear of honey, which allows them to be melted down into wax blocks using a solar wax extractor or other means. This method avoids the additional labour of washing the cappings.

An old Mountain Grey heather press loaded with a scrim bag of heather honey comb ready for pressing. [Brian Nellist]

The comb is pressed and honey flows from the outlet in the base of the press [Brian Nellist]

Commercial operators have sophisticated equipment for processing large amounts of heather honey. Smaller producers often opt for modern equivalents of the Mountain Grey (MG) or the Peebles press which meet current food hygiene standards. These provide a reasonable level of efficiency for a small-to-medium-sized operation.

TREATING HEATHER HONEY AFTER EXTRACTION

Heather honey naturally harbours wild yeasts. These yeasts need to be destroyed by a simple method of pasteurisation as a means of ensuring the bulk stored honey will remain in pristine condition. The pasteurisation process must not raise hydroxymethyfurfural (HMF) beyond the accepted level. Local authority trading standards officials are known to use sophisticated equipment to check HMF levels within honey offered for sale. Destruction of the yeasts can be achieved by taking the bulk honey and gently heating it for one hour at 60 °C. Thereafter, the honey is stored in cool conditions. Food grade plastic buckets with snap-on lids should be used to ensure that the honey does not absorb moisture (honey is hygroscopic). Storage at lower temperatures reduces the risk of fermentation.

A simple warming box can be made using a well-insulated cabinet, suitably sized to hold the honey container and a heat source. While sophisticated purpose-made warming cabinets with accurate temperature control are available, a simple DIY cabinet can be made using a discarded refrigerator. The heat source can be a thermostatically controlled warming cable specifically made for warming honey, such as that

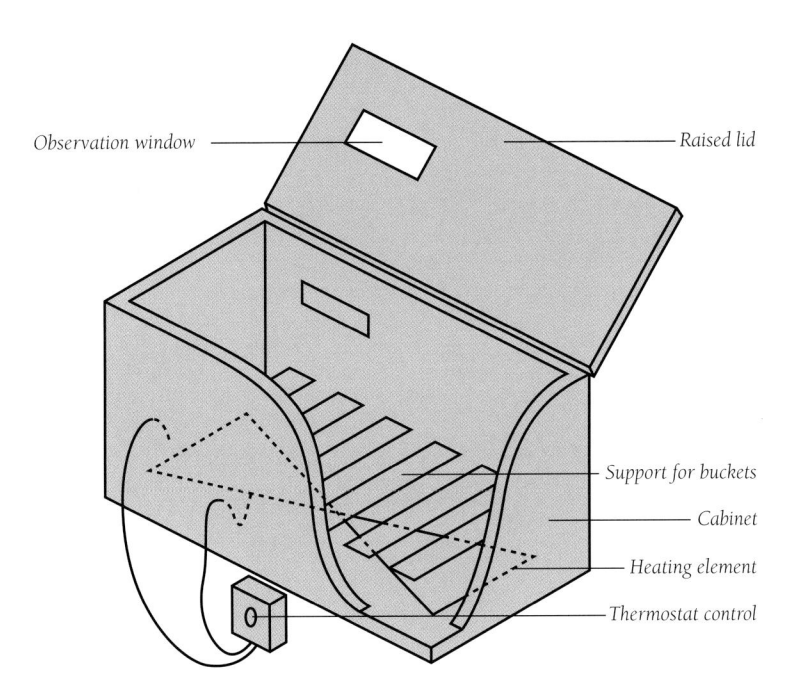

A thermostatically controlled warming cabinet

made by Ecostat. The cabinet should be of sufficient size to accommodate two ten-litre buckets. The buckets will need to be situated on a slatted floor with the heating element placed below but suspended above the floor of the cabinet to avoid contact with the cabinet structure to prevent damage and avoid risk of fire. A larger cabinet for an additional two buckets will require an additional 100 W element to maintain the heat required. A heat sensor is placed about 150 mm above the heating element, although some degree of positioning may be required to establish the optimum temperature control. A small double-glazed observation port can be incorporated in the lid or door of the cabinet to enable a thermometer to be seen, thus avoiding loss of heat to read the temperature.

A makeshift warming device for frames of comb containing honey. The heat from a tungsten light bulb is dissipated by a box of galvanised steel sheet. This provides gentle heat below frames contained in a super above [Brian Nellist]

The warming device with combs of honey in a super above it [Brian Nellist]

Methods Used to Obtain and Process Heather Honey

The user may need to trial the warming cabinet to get optimum results.

It is essential to understand that honey deteriorates when heated as the natural enzymes present are easily destroyed and chemical changes take place as the volatile elements present are driven off. Heating also raises the HMF level. Use minimal heat in processing the honey to avoid damaging it.

As previously stated, some beekeepers warm combs prior to extraction and, in my experience, I have found that this substantially increases the problem of frothing. In my opinion, it is folly to bottle ling heather honey straight into jars, unless it is for immediate use. Ling heather honey which has not been heat treated is highly likely to ferment if it is stored in jars because of its naturally high water content and the presence of natural wild yeasts.

BLENDING LING HEATHER HONEY

Ling heather honey is an ideal honey for blending with other honeys. Heather honey is the ideal medium for enriching bulk-stored oilseed rape honey which, to many, may otherwise taste somewhat bland. Heather honey is generally added at the ratio of 4–5 lb heather honey to 28 lb of oilseed rape honey. The oilseed rape honey (assuming it was strained prior to storing in bulk) is gently warmed to a point where, when stirred, it resmbles a thin gruel but is not fully liquefied. The heather honey should be warmed to the same temperature. This enables the blending process to work well with a resulting product that is uniformly mixed. The added heather honey is enveloped carefully into the bulk by stirring backwards and forwards in a sort of paddling

Blended heather honey. Note the rich colour [Brian Nellist]

motion, ensuring no air is added to the mass. The process of blending can take several minutes. It is recommended that the blending operation is carried out when the honey is intended for immediate use. It produces a soft-set honey with a surreal flavour. It is easy to remove from the jar and spreads with ease. It is an epicureal delight.

HEATHER SECTION HONEY

Section honey is a rarity in the British Isles. The sections are held in a section crate that is, in essence, a bottomless box constructed to hold from 16 to 24 sections, depending upon the size and type of crate chosen. The largest size section crates are rarely successful in the British Isles because of the climate and the lack of a sustained heather nectar flow because of the fickle weather conditions.

Methods Used to Obtain and Process Heather Honey

Heather honey is a very difficult crop to obtain at the best of times. The most experienced heathermen will concede that heather honey section production takes a number of years to master. The key difficulties are the requirement for continuous good weather for an extended continuous nectar flow and the fluctuations in the seasons that make it difficult to forecast when the heather will begin to yield. In addition, only exceptionally strong colonies will yield good quality sections.

Some strains of bees are loath to work sections. I have found that there are some particular strains that will work them, while others will not, irrespective of measures to coax the bees to enter the sections. Only a strong colony will work on sections, so no effort should be spared to make the sections attractive to the bees. Bees are gregarious insects and the single, unbroken cluster is how they naturally organise themselves. It is contrary to this distinctive habit to be divided by little boxes.

It is important to understand that for sections to be successful, it is necessary to use a colony at maximum strength. This means much more than having the hive 'boiling over' with bees in time for the heather honey flow; it means having the greatest possible number of bees at the right stage of their physical development. Under normal conditions, bees contribute nothing in the way of nectar to the colony's stores until they are at least six weeks old (referred to in my youth as the forty-day rule). Bees for the heather will be developed from eggs laid by the queen at the end of June and into early July. As explained previously, this is a most important criteria for success.

The Use of Hanging Section Crates or Section Frames

Hanging crates or section frames are devices that hold several sections in line. They can only be used for the production of one tier of sections. The hanging frame is placed in the centre of the super box, between other frames and over the brood nest.

One of my great uncles was a great section beekeeper. Part of his management system was to use hanging crates specifically for early use in the season, above the brood nest with 'bait' sections as a means to get the bees to occupy the new ones. Of course, this was three-quarters of a century ago, in rural Warwickshire, in fields of pasture of old Kentish white clover (*Trifolium repens*). These acres of farmland were relatively untouched until the coming of the first of the two world wars. This pastureland for section honey was manna from heaven, until the pastures were poughed, never to be returned to their pre-war state owing to agricultural intensification.

To obtain section honey, I have found that the simplest and most successful method is the use of specially made section crates. The purpose-made crate is placed centrally over the brood box within an empty super. The void is packed with hessian sacking or foam as a means of retaining the heat given off by the brood nest below. A crownboard covers the arrangement to assist with heat retention. This is best covered with a 25 mm (1 in) deep insulation board.

Difficulties with Getting Bees to Work Sections

Experienced heathermen have found that bees need to be coaxed or enticed into section crates. This can be done by painting a thin layer of molten beeswax onto the basswood sections to give off a pleasing odour. In addition, it is usual to put a pair of partially-filled bait sections into the centre of the crate. Unfinished sections from previous seasons can be retained specifically for this purpose.

My late father was convinced that drawn comb sections encouraged the bees to occupy the section crate because of the scent of heather honey, coupled with foraging activity on the moor.

The Use of Specially Made Section Crates

My own preference is to use a specially made section crate that is fitted inside a standard National super box. This holds 18 sections and is placed centrally over the brood nest, ensuring maximum temperature from below. Good packing surrounding the section crate provides much needed extra insulation to keep in the warmth.

Sections are provided with two or four bee-ways, which describes the number of directions from which bees can access the section. To ensure that the comb faces are drawn flat, thin metal tinplate (or plastic) dividers are inserted between adjacent comb faces. These dividers have an integral bee-way. I found by chance that the use of specially made wooden dividers between the rows of sections is advantageous as the use of metal is somewhat alien to the bees.

Three partially-drawn sections are used to encourage the bees to work the new sections to be added to the contents of the crate [David Pearce]

A complete section crate ready to be placed on the hive [David Pearce]

Methods Used to Obtain and Process Heather Honey

Foam or hessian packing can be used to retain warmth within a super for section production or, as here, for cut comb [Joanne Badger]

Section boxes are usually 115 x 115 mm (4.5 x 4.5 in) made from American basswood, with the introduction of European lime wood in the last decade. My preference is to use the former; it is usually knot free and works so much better. The use of thin foundation is the norm; bees are reluctant to work thick foundation and it is less palatable when the comb honey is eaten. Unfortunately, it warps and buckles in changing temperatures. The use of medium thickness is probably better. The experienced beekeeper usually makes up sections for immediate use to avoid this problem occurring.

A method for obtaining perfect sections is by means of top and bottom starter strips of foundation cut from thin foundation section squares. The top piece is cut so that it reaches to within 10–12 mm of the bottom of the section. The bottom piece is cut to a height of about 6–7 mm. Each of the two pieces of foundation is held in place with a little melted beeswax. This arrangement leaves a gap of around 3–6 mm between the two pieces of foundation which overcomes the problem of warping and buckling mentioned above. Another advantage of this method of preparation is that the comb will be built to the bottom of the section with a minimum number of 'pop holes'. It should be noted that the foundation must be fitted in the section frame the correct way up, ie, with the points of the hexagonal cells at the top and bottom, as the bees build comb naturally. It is impossible to get perfect sections using short top starters that require the bees to generate the heat necessary to produce a considerable amount of wax. Large areas of drawn comb will inevitably be of drone cells which is regarded as inferior in appearance to worker comb.

Of the two kinds of sections – two bee-way and four bee-way – which type you use is a question of preference and availability. The top of the constructed section is split lengthwise with a 45° bevelled cut. The three other sides are either grooved or not grooved. The purpose of the grooves is to ensure the foundation is kept central in the section frame. V-shaped transverse grooves between the four sides permit the flat section to be folded into a square frame. The last corner is fastened with a lock joint. It will be seen that very little wood is left at the v-grooves and care is necessary to fold the sections without breaking them. A purpose-made jig simplifies matters when assembling a section. Sections are most easily made up if the timber is dampened so that the joints are less brittle. A warm damp atmosphere helps, as does covering the sections with a damp cloth and damping the joints with warm water.

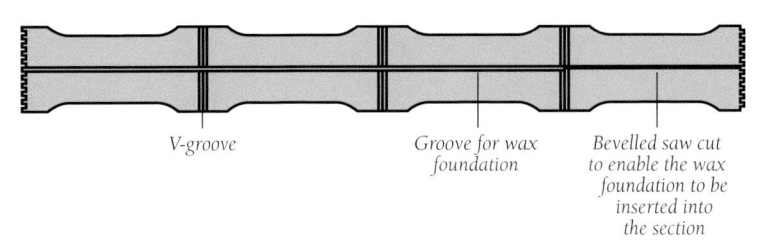

Methods Used to Obtain and Process Heather Honey

V-groove   *Groove for wax foundation*   *Bevelled saw cut to enable the wax foundation to be inserted into the section*

A four-way section prior to folding

A specially constructed handmade jig to assist with folding sections. The internal measurements should match the dimensions of the sections being used

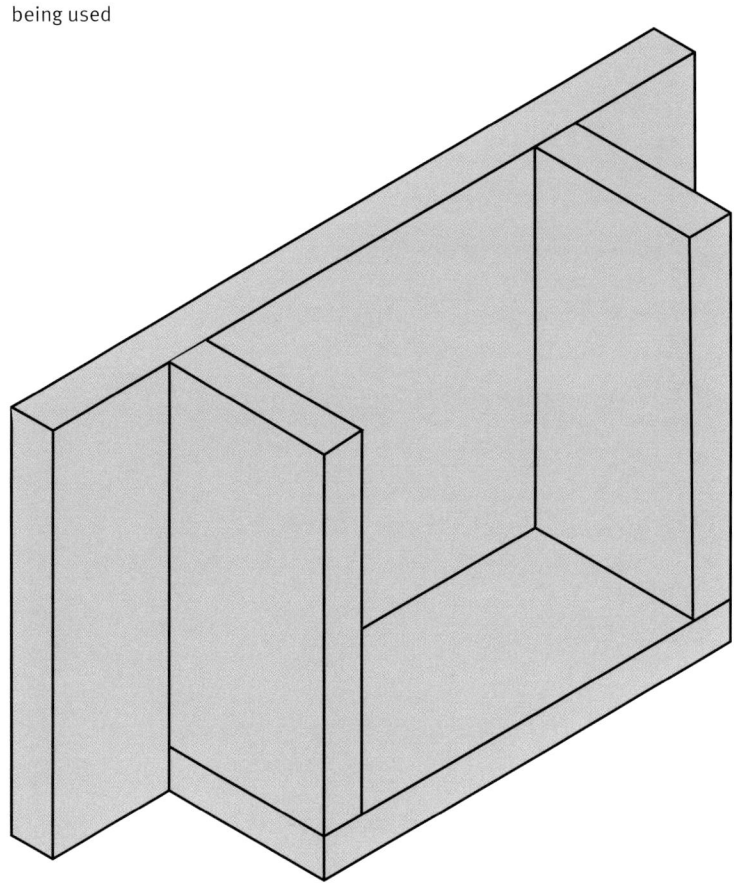

Sections being made up. Note the brush and water to dampen the joints and the folding jig in use [Hilary Sainsbury]

Section rack (cutaway) plus sections and dividers

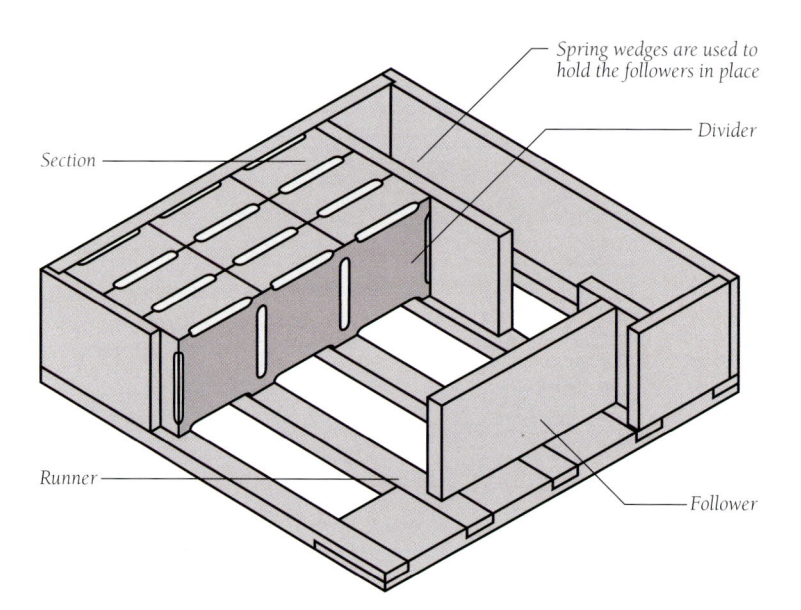

Spring wedges are used to hold the followers in place

Divider

Section

Runner

Follower

Separators placed between each row of sections prevent the bees securing them to each other with beeswax and propolis. They also act as guides so that the section comb faces are even and the bees are prevented from drawing cells too deeply, greater than the depth of the sections. The separators, like the sections, are constructed with two or four bee-ways.

Aftercare of Finished Sections

Completed sections should be removed from the racks with extreme care. Any adhering brace comb or propolis should be removed carefully with a sharp knife to avoid the scrapings dropping onto the face of the comb; this is most important for sections destined for the show bench. To finish off, the use of proprietary cellophane wrapping protects the comb face and the addition of a complementary label can make the product really attractive to the eye.

There is an understandable tendency for new beekeepers to use cut comb or Ross Rounds, the virtues of which are explained below. To quote my late mother in respect of such challenges: 'A faint heart never won a fair maiden or manfriend'.

ROSS ROUNDS OR COBANA CIRCULAR SECTIONS

The use of plastic circular sections is commonplace in the United States of America and Australasia and, to a leeser degree, in the British Isles, though there is a growing trend for their use. It is believed that the first circular sections were used in the USA in 1888 with more than a modicum of success. The success of these round sections was reported in an 1889 edition of *Gleanings in Bee Culture*. It is known that an English beekeeper, T Bloomer Cooper, trialled circular glass rings cut from round jars, secured to a frame holder. His experiments were undertaken a few years prior to the American trials but seem to have fallen by the wayside.

I am not sure or convinced that Ukranian beekeeper Petro Prokopovych (1775–1850) was the inventor of section honey as is sometimes documented. According to the late Colin Weightman, a Northumberland beekeeper, Robert Reed, was known to work several hundred skep colonies in Acklington and Morpeth from 1760 to 1812 and was producing sections. I visited Kiev, capital of Ukraine, in 1975. The museums there indicated that the nation had invented everything to do with modern beekeeping, emphasising that the country was the cradle of such, which is certainly an overstatement.

The hit and miss harvest of comb honey sections in the British Isles has led to the decline in their production, albeit the demand for them has never abated. Their use in heather honey production still holds an attraction with die-hard heathermen, although a good many have changed their production methods to cut comb or circular sections.

Initial Outlay for Circular Sections

The initial cost outlay for circular section equipment is relatively high. Despite the costs they do have an advantage over the original square sections in that the bees draw the comb fully to the outside of the ring, whereas the corners of square sections are often neglected. In addition, there is little to no waste with round sections compared with square sections and stale foundation can be removed readily and replaced with relative ease. Completed round sections do look very attractive.

A major advantage is that cleaning round plastic sections is accomplished more easily than the traditional wooden square sections. Circular sections are robust compared with the relatively thin lime-wood timber sections.

As with traditional sections, the use of bait combs is essential.

Management Techniques

The management of colonies for circular sections requires similar techniques to those for traditional section honey production. In my tour of commercial beekeepers in the USA in 1979, I was intrigued with their success with comb honey production. Mainly this is due to the bees they use, which have been specifically bred to work circular sections, in addition to good weather, ample forage and a more stable climate than in the British Isles. All these factors made it all much more of a successful enterprise.

The stocks to be used should be managed as discussed previously for traditional section honey. The bee stocks in the USA were managed by placing the circular section comb box directly over the single brood box (which was invariably Dadant or Langstroth) without a queen excluder in between.

Other producers I met added a super of drawn combs first, then added the crate of circular sections above once the bees had occupied the super and started to store nectar in it. This super was useful as a means to thwart swarming as it stopped the brood nest becoming congested with incoming

Crates of Ross Rounds being readied for use [Lester Quayle]

nectar. I have found that once the bees start to draw the foundation in circular sections it is good practice to reverse the position of the super of frames with the crate of circular section frames, placing them directly over the brood nest without a queen excluder. All in all, the beekeeper needs to trial a method and find that best suited to his or her own needs.

Removing the Circular Sections When Filled

It is good practice to reposition partially-completed and completed circular sections to the periphery of the section crate to ensure uniformity when finishing. When the time arrives to remove the completed circular sections, the holding springs are released to relieve the tension on the frame inserts.

Completed Ross Round sections [Peter Schollick]

To ensure maximum cleanliness and minimum scuffing to the comb faces, it is essential that the frames of finished sections are placed on a clean flat surface. Carefully inserting a sharp knife between the two half-frames, the two faces are prised apart with minimum force to release the finished round sections. The surfaces are cleaned up. Excess foundation is removed and covers are put in place. A wrap-around label printed with the producer's details, the product weight and any other legal requirements holds the covers in place.

CUT COMB

Cut-comb honey production has, to a large extent, superseded traditional square-section comb honey production. The purpose-made containers with snap-on lids ensure that cut-comb honey is kept clean and they keep insects at bay, too. The containers are easy to clean and pack. The lids allow labels to be fixed to the lid surface, making packaged cut comb honey very pleasing to the eye.

It is a common fault when packing pieces of cut comb that the honey from those cells which are cut through is not allowed to drain before packing the pieces. This results in packed comb surrounded by an excessive quantity of liquid honey.

Production Methods for Cut-comb Honey

All that has been said elsewhere in the text regarding production techniques and preparation of the colonies applies equally to cut-comb heather honey production. Working for ling heather cut comb requires the use of a purpose-built eke (or cog, as it is known in Scotland). This is a box that can be made from any timber that is absent of shakes or warping. The box should be robustly built, free of imperfections to minimise heat loss, relatively watertight and always bee-tight so as not to allow robbing.

A standard super is regarded by experienced heathermen to be too deep for bees to work successfully because of the relatively short honey flow experienced on the heather moors. The eke is a shallower version of the standard super, with a maximum depth of 80 mm. It is constructed to use a specially

Methods Used to Obtain and Process Heather Honey

A purpose-made eke with wax starter strips for cut-comb heather honey
[David Pearce]

adapted frame that uses thin unwired foundation. Beekeepers often cut horizontal strips of the foundation to a depth of 10–12 mm. The use of full sheets of foundation is not advised as they tend to buckle when being drawn; my colleague, Ivor Flatman, uses pieces of thin foundation cut in half diagonally that run the full width of the frame. The direction of the diagonal cut is alternated in adjacent frames. Narrow strips of wax foundation are secured to the frame using molten beeswax which is run judiciously along the abutment of the frame and the wax. Preparing starter strips should be done in a warm room to prevent the brittleness that occurs in cold conditions. Once the strips of starter foundation are secured to the timber frame, the bees will draw them with both drone and worker comb. The drop in overnight temperature and the generally cooler conditions on the moor see the bees taking

A polythene sheet is used instead of a queen excluder for production of cut-comb honey. Note the 25 mm (1 in) gap around the perimeter [Joanne Badger]

several days to draw the wax comb, more especially if there is a prolonged cold spell before the bees are able to get to work.

It is essential that a single sheet of horicultural-quality polythene is placed on top of the brood box combs as a means of keeping the drawn comb in the eke separate from the brood combs below. The bees have an aversion to plastic and will draw combs to leave a bee space above the plastic sheeting. The sheeting is made to cover the brood frame area with a 25 mm border on all four sides, thus allowing the bees sufficient access to the eke above. The gap allows free movement but the sheet acts as a barrier to the queen. The storage of pollen in the eke is most unlikely with heather-going stocks. Over the years I have not experienced any difficulty in getting bees to enter ekes. To ensure bees can be coaxed into an eke, beeswax can be painted

This comb of honey has been drawn from a starter strip of worker foundation. The bees have built drone comb in the lower half of the frame [Rupert Palmer]

liberally on the frames to give them a natural aroma and to entice the bees to work them.

The Cut-comb Container

Purpose-made cut-comb containers are constructed from food grade plastic that is both clean and hygienic and displays the contents to perfection. Each container will easily contain a 227 g (8 oz) piece of cut comb. The whole comb is placed on a secure surface and the finished area of comb cut carefully from the frame with a sharp knife. The comb piece is then cut into portions sized to fit the container. There are devices that allow pieces to be cut easily to a correct uniform size. The container is completed with a clear snap-on transparent cover.

Because cells are damaged in processing, each piece of cut comb is allowed to drain on a surface-mounted cake grid for

Cutting pieces of comb for packing in a purpose-designed container
[Willie Robson]

Once cut and drained, the cut-comb pieces are placed in their containers
[Brian Nellist]

the honey to drain into a tray. The slices of cut comb should not be placed into the cut-comb container immediately as the residual honey that drains from the severed comb will fill the container with liquid honey. The cut comb will be swimming in drained honey that makes the finished product unattractive and messy. Honey judges often encounter this fault when judging cut-comb classes. As part of the judging formalities, judges are empowered to remove the contents from the container to ensure the exhibit is true to both sides of the comb face. Needless to say, exhibits are marked down severely if the cut comb is found in such a condition.

Storage of All Types of Finished Sections or Cut Comb

The finished product in its container should be stored in a cold situation to reduce granulation and to ensure it is kept clean and wholesome. Storage in a redundant chest freezer is ideal. The temperature is set at 0 °C. This method will keep vermin at bay and will protect the finished goods until ready for use. It is advisable not to use a lower temperature in a deep freezer cabinet as the severe cold can and does crack the face of the comb. This is especially so when a large surface area of comb is packed.

THE RETURN FROM THE HEATHER MOORS

Having returned to the home apiary, the bees should be released without delay. The following day, the colonies can be prepared for winter. It is rare that the brood chambers of the returning colonies are not well provisioned with stores of both honey and pollen. These colonies are, above all, invariably very short of bees as little, if any, brood is reared while the bees are on the heather. Conditions on the moor ensure that there is little that can be done to encourage the bees to continue rearing brood. Once the colonies are back in the home apiary the beekeeper can address the issues of strengthening colonies for the approaching winter.

STRENGTHENING COLONIES FOR WINTER

It is particularly important to increase the numbers of young bees that will take the colony through the winter months to the next year. The best remedy is to unite a nucleus (if available) to each of the returned colonies; these colonies will overwinter splendidly. The arrival of new bees which, in the main, will be young bees, puts new heart into the colony. These bees will work the last remaining flora for pollen and nectar. There will be more young bees within a month, ensuring the colony has the best packing for winter – bees and young bees too. Colonies in apiaries that are near to watercourses may be readily supplemented by nectar from Himalayan balsam that will be available until the first frosts. Ivy nectar will also be widely available.

Uniting is best done by removing the older queen from the nucleus (if she came originally from the heather stock, now headed with a daughter or replacement queen), removing empty frames from the main stock and introducing the combs of bees from the nucleus directly into the parent stock.

Another common method of uniting is the newspaper method. This is best carried out during the middle of the day when the bees are flying well. The roof and crownboard are removed from the main colony. A sheet of newspaper that will cover the whole of the top of the brood box is pricked with small holes and placed on top of the brood chamber; it can be held in place with a temporary queen excluder if necessary. The contents of the nucleus (frames and bees) are placed in an empty brood box on top and to the rear of the brood chamber so that they are well away from returning foraging bees. The bees will chew through and remove the paper from the hive and this slows down the process of uniting, which ensures success.

Experienced beekeepers often dispense with the newspaper as they consider the nucleus bees are, in the main, young bees

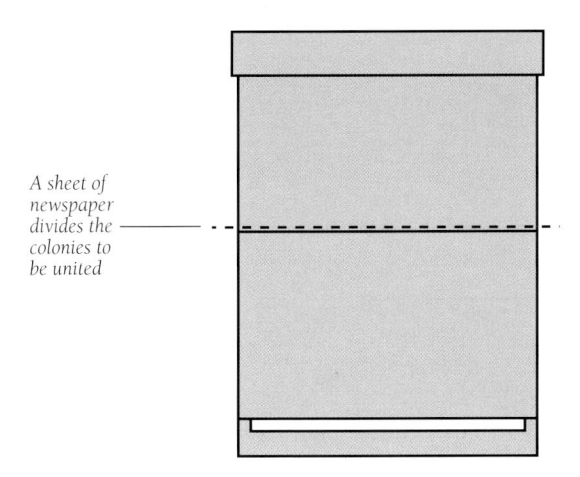

Uniting a nucleus to a colony using the newspaper method

A sheet of newspaper divides the colonies to be united

which will integrate without trouble. The newspaper method removes any risk of trouble and ensures the safety of the queen. New beekeepers are recommended to unite their bees over paper.

Understandably, it is not always feasible to have a spare nucleus colony available. In such situations, the beekeeper should place a brood chamber or super with frames of drawn comb below the brood box and directly onto the floor board of the recently returned colony. The surplus crop in the supers having been removed, a rapid feed of 4.5 litres (1 gallon) of thick syrup is given to the bees. The higher ambient temperature at the home apiary, additional room and available feed gets the bees into good heart and they will begin rearing brood again, although just for a few weeks. The additional young bees assist heather colonies to overwinter superbly.

In addition, unsealed honey is hygroscopic and readily absorbs moisture from the surrounding environment. The syrup feed allows the bees to seal any remaining unsealed stores of heather honey in the brood chamber, thereby reducing the potential for fermentation and dysentery later in the winter period.

It is essential that heather colonies are not overwintered on a brood chamber of stores, especially if this is National or Smith size; other brood chambers have the required depth for good wintering. A deep or, as a last resort, shallow box, preferably with combs, should be placed upon the floor, beneath the brood chamber with its sealed stores. This configuration is very much how a wild colony lives. It allows the bees to cluster as they would in a feral situation and they winter superbly. Bees do not overwinter well when forced to cluster on sealed stores. Bees in their natural state cluster below their stores and move upwards as the consumption of stores progresses.

The late Teddy Sonley, a lovely heather beekeeper from Kirby Moorside, North Yorkshire, listened to a lecture I gave in the 1980s on managing heather bees. Teddy remarked that wintering bees in this manner sounded good advice. The following April he contacted me to advise that having applied my advice to his own colonies, he had carried out an artificial swarm a good four weeks earlier in the season than in previous years.

BEES IN THE FERAL STATE

The illustration shows the natural state of a honey bee nest in the wild. The heather beekeeper is encouraged to overwinter

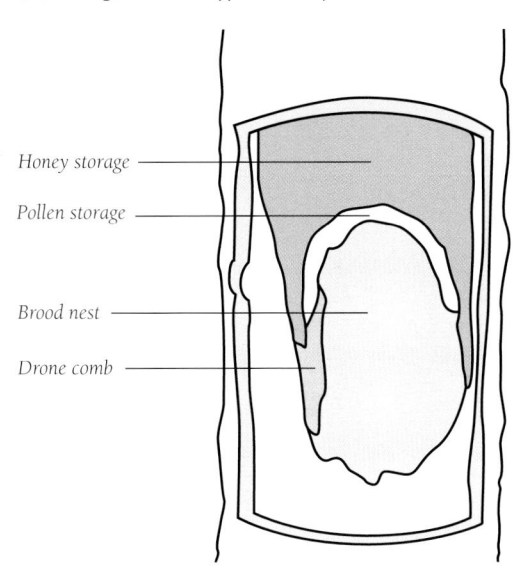

The arrangement of a typical honey bee nest inside a cavity

Honey storage

Pollen storage

Brood nest

Drone comb

his or her bees on this principle. It bodes well for overwintering by non-heather bee stocks, too; this is how bees live in nature.

The natural juxtaposition of stored honey, pollen and brood is shown and it can be seen that wild colonies do not overwinter on sealed stores but hang or cluster below them and 'eat' their way upwards onto the sealed comb. It is well known that (prior to the arrival of the varroa mite) wild colonies lived in the same location year after year, only dying out if they swarmed late and cast themselves to death. In this event, they would be replaced by incoming swarms in the following year.

I was very gratified in 2009 when Professor Tom Seeley stayed with me and I explained my theory on overwintering. I was pleased to learn that such research was being undertaken at Cornell University, Ithaca, New York. The success of the natural way was demonstrated by Professor Seeley at The Eastern Apicultural Society Conference, 2012. Seeley's researchers noted that feral colonies they observed fared better than those colonies that were overwintered and managed on open-mesh floors.

For several years, I have noted that my own nucleus colonies overwinter better when kept on traditional solid floors rather than mesh floors open to the elements. Ian Craig, a well-respected Scottish beekeeper, told me that his own colonies in Renfrewshire were better for overwintering on traditional floors than on open-mesh floors; the number of losses in spring were noticeably less, as in my own situation. The consideration that honey bee colonies in nature generally do not have an air draught from below probably gives credence to this mode of overwintering.

Beekeepers that overwinter their bees in Commercial or other deep hives will have the required depth of comb. In comparison, National, Smith or WBC brood chambers are of a much reduced depth.

Some beekeepers use a super rather than a second brood box as a measure to overcome the weight and cumbersome nature of a double deep brood chamber hive. In any event, the addition of either a brood box or super is best regarded as a temporary restriction to hive manageability. The benefits outweigh the disadvantages.

POLLEN-CLOGGED SUPERS

Beekeepers who use the smaller types of brood chamber (National, Smith and WBC) and persist with single brood chamber management may notice that the combs in the super placed immediately above the queen excluder become full of pollen. Closer observation reveals that this is a continuation of the natural pollen arch through the queen excluder above the brood. There are strains of bees that will, if permitted, surround the brood with pollen more or less a full 360 degrees. I saw this first hand with the late Bill Bielby in 1973, at Fountains Abbey, North Yorkshire, in colonies of the black native bee *Apis mellifera mellifera* that was identified through wing venation indexing.

The term 'pollen-clogged' refers only to combs containing pollen that the bees cannot use because it is stale and has dried out. Surplus combs of stores of pollen and honey within the brood chamber should be removed in the active season, when the colony is expanding. Under normal conditions, a colony only needs a comb of stores on each flank of the brood nest. The expanding colony needs good comb with space available for breeding purposes to ensure a strong foraging force at the appropriate time to gather a surplus.

IF IN DOUBT, DON'T

Once the bee stocks have been prepared for winter and the beekeeper has undertaken all the required operations to get the bees in the best state of readiness for the coming season, they should be left alone as the need for further manipulations has passed. Mouseguards should be fixed in position and, if entrance blocks are used, they should be inverted so that the entrance does not become blocked by debris accumulating on the hive floor.

It is good insurance to surround hives with small-mesh chicken wire or black plastic as a deterrent to woodpeckers in severe winter conditions. These materials make it difficult for these predators to gain a foothold in order to attack the walls of the hive.

The activities of the bees before winter finally sets in can tell you many things. If you see the bees taking pollen into the hive it is a sign that the queen is still in lay. Bulbous pollen loads are carried by younger bees, whereas light loads are carried by older ones. The presence of drones is a sign that the

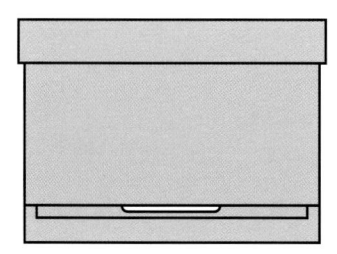

The entrance block in an inverted position for overwintering

colony may be queenless or that there is a drone-laying queen or drone-laying workers.

In spring, wax fragments on the floor near the entrance (like sawdust) are evidence that the bees have moved to new areas of the brood chamber; the debris is the removal of wax cappings on the stores.

Should you place your hand over the feed hole in the crownboard in the dead of winter and detect a high degree of heat from the cluster then this may be a sign of continued breeding, which may be an indication that the colony has nosema; the bees may increase brood production as a means of eradicating the malady. A sample from the colony should be analysed to check for infection as soon as is practicable. If found to be positive, a Bailey comb change can be performed as soon as conditions permit.

Observation by the beekeeper is an essential requirement of successful bee management; it is a prerequisite to understanding the colony without recourse to unnecessary manipulations in or out of season.

FROM HIVE TO MARKET: PRESENTATION

Those in the world of marketing are all too familiar with the phrase 'proper presentation speaks a thousand words'. In the 1970s, there was a rise in interest in more traditional food production and self-sufficiency, popularised by the BBC sitcom 'The Good Life', an entertaining smut-free situation comedy that brought rural activities and people's wellbeing to life through the antics of a young couple's dalliance with self-sufficiency while living in suburbia. The 1980s saw the rise of farmers' markets as a means to buy and sell local produce. These markets provide an alternative to other food shops and are usually held at a weekend in cities and towns alike, very often in close proximity to the shopping centre.

Daily television programmes emphasise how much interest there is in food, its cooking and its provenance, over and above the simple necessity of sustenance. Yet it is obvious to me watching consumers purchasing large quantities of ready-made meals, that there is still a way to go. Fortunately, the tide is turning and there is rising interest in purchasing good quality, high value foods. Long may it continue.

The public clamour and hostility to genetically modified (GM) food crops and neonictonoid pesticides has reinforced the need for reassurance that foods are wholesome and produced both ethically and sustainably. In the past 20 years, there have been revelations about 'mad cow' disease and, more recently, the contamination of meat products in the food chain. These do little to reinforce public confidence or challenge the view that the food we eat is little more

Attractively presented jars of honey and cut comb for sale at Chain Bridge Honey Farm [Hilary Badger]

than 'conveyor belt' food that is intensively mass produced and unwholesome.

A good many of the honey-consuming public see honey as a wholesome natural product, produced by bees and harvested by man. Heather honey comes into its own as it is sourced from non-farmed heather moorland that is not interfered with at all other than by traditional moorland management that is generally chemical free.

Over the years I have found that good quality honey will sell itself, always provided it is reasonably priced, marketed attractively and that the buyers have confidence in the source of the product and its origins.

Those selling honey should ensure they are aware of the relevant legislation.

BOTTLED HONEY

A trend that has gathered pace these past 20 years is for beekeepers to abandon the traditional squat 1 lb honey jar for the 340 g (12 oz) hexagonal honey jar which is visually of similar size. The food industry carefully chooses packaging sizes as a means of enhancing profit margins by charging more for less to an unsuspecting public. One could say that beekeepers have latched on to the concept too, and why not?

Fine samples of ling heather honey [Brian Nellist]

An attractive label is an important aspect of presentation [Brian Nellist]

YORKSHIRE
HEATHER HONEY
from B. & M. Nellist
Mill Cottage
Egton Bridge
Whitby
454 g   (1lb.)
Produced in U.K.
Best before end   Lot No.

Commercial beekeeper, John Home, bottling honey, Fosse Way Honey, Warwickshire [Richard Stanton]

Using a filling machine speeds up the process considerably [Richard Stanton]

LING HEATHER HONEY FOR EXHIBITION

Ling heather honey is the simplest of all honeys to prepare for show and exhibition, whether for a local produce show, a major agricultural show, or the National Honey Show.

METHOD

Ling heather honey will flow more readily if it is warmed thoroughly first, but (and an essential but) it should not be overheated, which can so easily happen. A simple method for warming the honey is to set a 40 W tungsten light bulb in a brood box to act as the heating medium. As these can now be difficult to obtain, a warming cabinet may be used as an alternative.

Place a super of combs of heather honey above a warming box containing the light bulb or other heat source. This should be partially covered to protect it from dripping honey and to dissipate heat. Both of these boxes should be well covered to retain the heat.

The honey should be heated for no more than six to eight hours. A good heather honey judge can, and will, detect any hint of caramelisation through overheating. Better to check regularly and heat only for the minimum time necessary.

Line a large basin with a single sheet of coarse muslin or linen scrim to act as a straining cloth. Scrape the warmed heather honey from the comb into the cloth-lined basin, ensuring no pollen, mixed flower honey that might be present, or any detritus is included. Examining a frame in bright daylight will

The even golden colour of backlit heather honey in the comb. Note the absence of any pollen [Brian Nellist]

show up the different shades of any different honeys within the comb. These should not be mixed with the ling. If they are it will make the honey a blended heather honey.

The straining cloth can be secured at its corners to two heavy chairs, suspending it over the basin and allowing the honey to drip through the cloth. This process will introduce the large air bubbles required in exhibition heather honey.

Once the honey has filtered through, cover the basin to ensure nothing gets into the honey and place it somewhere warm. The space above the hot water cylinder of a large airing cupboard is ideal to allow the honey to gel for a minimum of five days.

The honey can be poured or spooned into jars, ensuring that large air bubbles remain present.

To ensure the honey is presented correctly for the honey judge, the exhibitor should take a little time preparing the glass

An open jar of ling heather honey showing the judge's strike mark made to check thixotropy, an indicator of purity [Brian Nellist]

jars before bottling. They should be washed in a dishwasher and then thoroughly dried and polished with a clean well-washed linen cloth. A well-worn linen handkerchief is ideal as all the lint has been washed off. Pay careful attention to the rounded edges at both the neck and base of the jar and make sure they are well polished. If a dishwasher is not available, wash the jars thoroughly and rinse in cold water; heat the jars in the oven at a very low temperature for 45 minutes. The jars will shine up well; a higher temperature can bake on a glare that can leave a bluish hue.

Prior to filling, jars should be matched in pairs and checked for any flaws in the glass. The two jars must be identical in all respects. Jars need to be filled to capacity if they are to be of regulation weight as there is a fair amount of air incorporated into the honey mass.

Filled jars with lids on should be stored in a totally dark, warm place. After a time, it is usual for fine granules to appear. These can be removed by standing the jars up to their necks in hot water, to a maximum of 55 °C. Leave until the granules disappear. Heating with a steamer is another way of removing these flint-like granules from the body of the honey. Care is essential to avoid overheating.

Three days prior to show time, carefully remove the original lids and replace them with new ones after cleaning up the jar tops and the screw threads. Do not change lids after arriving at the show, especially if judging is to take place shortly afterwards. Removing lids is unnecessary and it will release the much-required aroma that the judge is seeking. In addition, if the jars and lids are clean, an experienced judge knows whether any honey on the lid has been put there prior to travel or by careless handling by show stewards.

The experienced exhibitor makes a simple positioning guide collar so that he or she can ensure the class label is fixed at the correct height to the dimension stated in the schedule and that it is centred between the jar seams.

If you find the honey is not in a gel-like state but moves readily when the jar is tilted to one side, it is almost certainly a heather blend. Such honey, which is perfect in every respect, will easily be beaten in competition by a pure heather honey. Fortunately in the major honey shows there are classes for both pure ling heather and blended honeys.

The thixotropic property of ling heather honey has been mentioned elsewhere. Because of this characteristic, if an open jar containing a pure sample is inverted, the honey should not flow from the jar. By stirring part of the honey in the jar, it can be made liquid temporarily. Thus this portion of honey

Position a guide collar to ensure the consistent, correct height for class labels. These labels should also be positioned midway between the jar mould seams. The template height is adjusted according to the show schedule requirements [Damien Timms]

will flow like ordinary honey. After 24–36 hours it will return to its original state. Professor Pryce Jones found that this phenomenon was caused by certain proteins within the honey.

To check both a sample's purity and whether it has been heated, the honey judge runs a tasting rod across the surface several times. He or she will observe the sharpness of the edges between the disturbed and the undisturbed honey and watch the rate of collapse of the edge of the undisturbed honey closely. If this edge stays perfect, the sample is most likely to have been heated.

To protect exhibition honey from granulation, exhibitors often place it in a refrigerator or a domestic freezer. This

Heather Honey: A Comprehensive Guide

A backlit jar of ling heather honey showing perfect colour and distribution of air bubbles for exhibition purposes [Brian Nellist]

slows the process, but it also ages the honey and darkens the colour.

An experienced heather judge will tell you all that there is to know about samples of honey put before him while judging.

When I am judging ling heather honey, I always place a sample of the honey between my thumb and index finger and rub them together. This provides an indicator of the honey's purity as any granules of other honey can be detected. I was complimented on this practice by the late Bernard Leafe, a well-respected Yorkshire beekeeping exhibitor, while I was judging at the Stokesley Show, North Yorkshire. It is a good tip for all exhibitors to note. However, some judges rely on taste alone for this test.

Ling Heather Honey for Exhibition

These are some hints on exhibiting bottled heather honey.

- Read and understand the show schedule. Note the type and size of jar required, along with the lid and positioning of the class label. If in doubt, ask.
- After washing the honey jars, polish the insides, making sure that the bottom edges are polished.
- Select matching pairs of jars in all respects before filling them with honey. Beware of moulded text or numerical characters on the rounded edge at the base of each jar. These should also match.
- Fill the jars to the top and secure the contents with a lid. Remember, the honey is full of air and you must expect some settlement.
- A few days before the show, select good clean unmarked lids, preferably with built-in flow seals if metal lids are specified. Lids must match. Owing to the poor quality of metal lids, show schedules are increasingly allowing the use of plastic lids.
- Remove the old lids and clean up the jar threads and the top lip of the jar thoroughly. Secure new lids and store the jars in a cool place. The new lids should not be removed before judging. Do not be tempted to change lids when staging exhibits.
- A good judge will not penalise an exhibit that has honey on the inside of the lid. He or she knows that the staging steward may have done this, as most exhibitors are very careful when transporting their exhibits. The judge knows good and bad preparation immediately on sight.

- Clean any finger marks off the jar by polishing the outside carefully. Some exhibitors clean the glassware with methylated spirits; the odour quickly abates.
- Use a marking collar to place the class label midway between the jar mould seams to the correct height stated in the schedule.
- Finally, remember that the judge has been or still is an exhibitor.

The author judging heather honey at the Countryside Live show, Harrogate, 2013
[Wendy Maslin]

FINAL CONSIDERATIONS

It is hoped that the reader will have dispensed with any notion that going for heather honey is a pursuit fraught with obstacles, snags and pitfalls; none are insurmountable. It should be regarded as a great and enjoyable adventure to be undertaken by all (if possible) at some stage in their beekeeping careers. Mention has been made that the majority of books are not helpful in inspiring newcomers to visit the heather moors. My only suggestion is to try it, perhaps with just one colony at first. The past few years have been trying all round for practical beekeeping. Indeed, this past season, 2015, proved to have been the most challenging for over 25 years. Poor seasons and variable weather have been with us since time immemorial so, as they say in Yorkshire, 'give it a go'.

The quality of the honey and the surplus to be obtained is geared to many factors, namely:

- the nature of the soil where the heather is grown
- sites with over 100 acres of heather in order to obtain a surplus
- the correct altitude to site the hives
- the absence of rainfall as rain curtails foraging
- the age and condition of the heather. The best nectar yield is by plants that are no more than 300 mm high; large scraggy plants are generally worthless
- the best moors are those that are managed as driven grouse moors which are burnt every 8–12 years. Thereby the heather plant regenerates after two years with young flowers that yield large quantities of nectar

- sheep tending ensures that the heather plants are kept cropped down, creating flushes of young shoots
- site the hives in the heart of the flowering heather
- ensure the hives are securely strapped up to stop them being knocked over by sheep using them as scratching posts.

So, make the best of it. Take cognisance that the primary difficulty for obtaining heather honey is that we are forcing the bees into an unnatural mode of existence. To recap, in early spring through to early summer the compelling force throughout all nature is to increase and go forward; in late summer, with shortening day length, the urge is to ease off for winter rest. Bees are no exception to this phenomenon. The success at the heather moors therefore relies on the following factors:

- only very strong colonies will be successful at the moors. You need plenty of unsealed brood right up to the time the heather begins to flower
- small hives are best with a maximum of ten combs of brood
- place warm packing over and around the supers. Plenty of bees is the best packing for the moors
- ventilation is vital when transporting the bees to and from the moors. The ventilation is best given at the top and bottom but never at the entrance. Do not underestimate the importance of this. I have met many beekeepers who mock at this suggestion. Over the years I have come to the conclusion that they do not know what a strong colony of bees is really like
- finally, be respectful to the moor and its owners when using vehicles on the moor.

SECTION 2
SUGGESTED FURTHER READING

- Armitt, JH (1952). *Beekeeping for Recreation and Profit.* Birmingham, AA Ladbrook Limited.
- Atkinson, JH (1999). *Background to Bee Breeding.* Hebden Bridge, Northern Bee Books.
- Caron, DM (1999). *Honey Bee Biology and Beekeeping.* Cheshire, USA, WICWAS Press, LLC.
- Cooper, BA (1986). *The Honeybees of the British Isles.* Codnor, British Isles Bee Breeders' Association.
- Couston, R (1990). *Principles of Practical Beekeeping.* Mytholmroyd, Northern Bee Books.
- Deans, ASC (1963). *Beekeeping Techniques.* Edinburgh and London, Oliver & Boyd.
- Gilman, A (1928). *Practical Bee-Breeding.* London & New York, GP Putnam's Sons.
- Gregory, P (2013). *Healthy Bees are Happy Bees.* Stoneleigh, Bee Craft Limited.
- Hamilton, W (1951). *The Art of Beekeeping.* 3rd Edition. York, The Herald Publishers.
- Hooper, T (2008). *Guide to Bees and Honey.* 4th edition. Hebden Bridge, Northen Bee Books.
- Hooper, T and Morse, RA (eds) (1985). *The Illustrated Encyclopedia of Beekeeping.* Poole, Blandford Press Limited.
- Jefferson, T (2014). *A Practical Guide to Producing Heather Honey.* Hebden Bridge, Northen Bee Books.
- Mesquida, J (1954). *Elements of Genetics with Special Reference to the Bee.* Translated by Revd E Milner. Codnor, British Isles Bee Breeders' Association.

- Neighbour, A (1878). *The Apiary*. London, Kent and Co.
- Seeley, TD (1995). *The Wisdom of the Hive*. London, Cambridge, Massachusetts, Harvard University Press.
- Seeley, TD (1985). *Honeybee Ecology*. Princeton, Princeton University Press.
- Seeley, TD (2010). *Honeybee Democracy*. Princeton, Princeton University Press.
- Shannon, D (2015). *A Comprehensive Guide to Preparing Exhibits for a Honey Show*. Hebden Bridge, Northern Bee Books.
- Shimanuki, H, Flottum, K and Harman, A (2007). *The ABC & XYZ of Bee Culture*. 41st Edition. Medina, USA, The AI Root Company.
- Sims, D (1997). *Sixty Years with Bees*. Hebden Bridge, Northern Bee Books.
- Smith, J (2011). *Better Queens*. Nehawka, Nebraska, X-star Publishing Co.
- Taylor, EH (1906). *Bee Appliances and How to Use Them*. 3rd edition. St Albans, Gibbs & Balmforth Limited.
- Tinsley, J (1945). *Beekeeping Up-to-date*. Glasgow, Aird and Coghill Limited.
- Turnbull, B (2011). *The Bad Beekeepers Club*. London, Sphere.
- Whitehead, SB (1954). *Bees to the Heather*. London, Faber and Faber.
- Winston, ML (1987). *The Biology of the Honey Bee*. London, Cambridge, Massachusetts, Harvard University Press.

# SECTION 3:

HISTORICAL ASPECTS OF HEATHER AND OTHER FLORAL HONEY PRODUCTION

Bell heather [Brian Nellist]

A BRIEF HISTORY OF TAKING BEES TO THE HEATHER, FOLKLORE AND LITTLE-KNOWN FACTS

A history of migratory heather-going and the folklore around it is a topic that is rarely, if ever, aired in books. The following account is my own thoughts on events. As I was told in my schooldays, history is 'his story'. This is my interpretation of proceedings and might be construed as a possible opener to an interesting topic of conversation with fellow beekeeping enthusiasts. I suppose it could be said that my narrative is the opening shot to a worthy theme that may inspire others to carry out their own research which might unearth new information that has been lying in some museum or library vault gathering dust. Who knows?

The knowledge as to when beekeepers first took their bee colonies to the heather moors is really lost in the depths of time. There are few, if any, known early references to when the practice of heather-going originated, other than it seems to have been an ancient custom of those beekeepers who, by chance, were probably living in close proximity to the heather moorland. The earliest written record that I have found that chronicles the moving of bees to the heather is in a scarcely available little book by Jacob Isaacs, *The General Apiarian*. This was published in 1799 in Exeter and a revised version was published in 1803. Isaacs lived in Moretonhampstead, Devon. The following extract is from the 1803 edition [sic].

> *In a rich country, destitute of those trees which retain the sweet exudation, commonly called honey dew, the*

Bee-pasture is over, in general, in the month of June: but the heath, an almost never failing-source of honey, begins to blossom in general in July. About the middle of this month, therefore, carry your poor swarms into or near the heath, as well as stocks that have swarmed, but found light: and, if the weather be mild, or a little dry in August, they will soon fill the cottage-hive, out of which you may take at Michaelmas (3rd of the four quarter days – 29 September, the other-quarter days, being Lady day, 25 March, Mid-Summer Day 24 June and the fourth Christmas Day 25 December) when you bring them home, as much as to reduce the weight of it to twenty-four pounds, provided you take out no brood. I have proved this by repeated trials. I think they are not above a mile from the heath; and that if they are moved at all, it should be to the extent of three or more miles. I move mine six miles, and never find one come back until they are brought back in the hive at Michaelmas, in a rich and strong state.

They are carried by two men, on an elastic bier, formed of two round dried aller poles, 7' 00" long, united in the middle by a doubled canvas cloth, at twenty-one inches asunder and long enough to receive three hives. The bees are well stopped in, and cords are drawn about the floors and hive to prevent their sliding. I have sometimes carried them on a horse.

But let me charge you, when the hives are close and very full of Bees, always to stop the entrances with perforated tin plates. I have known a swarm, consisting of at least thirty thousand bees, destroyed in hot weather, for want of air, in a small hive, by being carried only two miles.

A Brief History, Folklore and Little-known Facts

The hives are also enclosed in separate sheets, for greater security, and when placed on the bier cloth, the poles come up several inches on their sides, and the hives are so tied so as not to strike in carrying. The poles are supported by web-strings from the shoulders of the carriers, and held by their hands, like a sedan chair. They can carry three swarms away, but on account of the increase of weight, the carriers can bring home but two at a time.

On the 30 July last, one poor swarm, which, with all that it could collect here for a month before, weighed only five pounds and four onces, was carried into the heath of Dartmoor, and brought back two months afterwards, with an increase of twenty-four pounds and a half. The increase of others were nearly proportionable. And I have reason to believe that all this increase was made in the month of August. Bees sometimes, in three weeks will gather more than a sufficiency for them for a whole winter. It is sometimes well for them that they can do it, otherwise, in some years, all the Bees in the country would die for want of the compassion and care of their owners.

If, when the hay is mowed, the pasture of your bees be gone – if no lime or sycamore trees or bramble bushes be in any number near – if no fields of late beans, buck-wheat, or white clover be within their reach – or if heath be at a distance, remove them to it; and if you do this discreetly, according to the foregoing hints, your trouble and expense will be amply repaid.

Having read the foregoing narrative, the reader will note that heather honey is not recommended for its own sake but merely as a means of strengthening and provisioning swarms and light stocks. There is no suggestion of sending strong hives to the moors at all, because heather honey was not wanted as a harvest. When Isaacs refers to the occasional possibility of taking surplus honey from stocks that have been brought back from the heather, he does so without any enthusiasm. It was not the object he had in mind when he sent his bees to the moor, especially with the elaborate means of transport.

Some 12 years later, Robert Huish (1777–1850, born in Nottingham, died in Camberwell, London) in his book *Bees: Their Natural History and General Management* tried to persuade (unsuccessfully) some of the Scottish beekeepers of Dingwall, Ross and Cromarty to move their bees to the heather; he found them disinclined.

It is recorded by John Wighton in 1842 that it was about this time that heather honey became truly appreciated. He states in his book *The History and Management of Bees* that in the south of Scotland bees are taken to the heather. Elsewhere he remarks: *Honey from wild flowers is considered the best, that from heath for example. The stone which the builders rejected is become the head of the corner.*

According to the late Colin Weightman, Robert Reed was thought to have been the first known commercial beekeeper in Northumberland, known to work several hundred skep colonies in Acklington and Morpeth from 1760 to 1812. Robert visited all the local fairs, marketing his honey and giving skep-making displays to the public. Although it is not recorded, it seems more than feasible that he was one of the early heathermen.

A Pettigrew, in his *Handy Book of Bees*, makes mention of heather honey being harvested before 1870, no doubt by skeppists. William Herrod-Hempsall, in his two-volume

A Brief History, Folklore and Little-known Facts

Bee-Keeping New and Old, described with pen and camera, refers to those beekeepers who were fortunate enough to have access to a horse-drawn cart; these were probably the first migratory heathermen. Over the years, I have met a number of beekeepers from Northumberland who were told that many country 'cottager' beekeepers were not averse to taking their hives to the moors by wheelbarrow. The late Albert Hind, northern correspondent of the *British Bee Journal*, kept me amused with stories of how his great-grandfather used a specially adapted wheelbarrow to take a couple of stocks from his apiary in Wooller to the nearby Northumbrian fells, a distance of over 20 km (12 miles). However, those beekeepers fortunate enough to live on the periphery of the moors were, without doubt, the earliest ones to harvest heather honey without the hassle of getting their stocks to a heather stance.

GERMAN HEATH BEEKEEPERS

Heather honey production was not only confined to Great Britain. The *British Bee Journal*, 15 January 1884, cites German beekeepers moving bees from Hanover to the Lüneburger Heide, a practice that had existed for a great number of years. Many of these beekeepers made their living from the extensive plains of heath and heather in this part of Germany. They are known as 'heath beekeepers' and it was not unknown for them to travel up to 65 km (40 miles) to the heathlands. Owing to the distances involved, the beekeeper would set off in the evening, after packing his hives onto the *Bienenwagen* (bee wagon); yet, very often, he was unable to get to the heathlands before daybreak. Should this happen, he took the hives off the wagon, set them a distance from the roadside and let the bees fly. After sunset, he would repack the hives and set off for the final leg of the journey. It was obviously profitable to undertake such an arduous venture.

MIGRATORY HEATHER BEEKEEPING FOR THE HUMBLEST BEEKEEPERS

In the late 1970s, the late Alf Hebden related to me that shortly after the First World War, beekeepers in Leeds were fortunate to get a lorry to take their bees to the moors. A well-known enterprising transport operator in Leeds, Sammy Ledgard, converted one of his lorries to provide 'charabanc' services throughout the summer weekends for the locals, enticed to go to nearby beauty spots in the Yorkshire Dales. During Wakes fortnight in early August, he would offer his lorry to beekeepers to take their hives to Denton Moor, Otley, West Yorkshire, and also to Pateley Bridge, further north. At that time, Mr Sainsbury, the secretary of the original Leeds Beekeepers' Association (1898–1938), arranged for anyone interested to have their bees packed and ready for picking up, each beekeeper being charged the princely sum of less than a shilling (5 decimal pence) for a maximum of four hives. Three weeks later, the bees were brought back. Another beekeeper travelled with the driver to supervise loading and unloading.

Brian Eade recollected that, as a schoolboy, he remembered his father, John (Eddy) Eade, a police officer and semi-commercial beekeeper (later to take over Mountain Grey Apiaries in the 1960s), hired a local transport operator each year to take more than 80 hives to Saltersgate Moor, North Yorkshire. He recalled vividly the return journey following a bumper crop year. The lorry, loaded with hives with full supers could not climb the B1248 hill at North Grimston. The driver had to put chocks under the rear wheels. Twenty stocks were unloaded to allow the vehicle to get to the hill summit and it returned much later that evening to collect these colonies, once the rest had been off-loaded at the home apiary at Bubwith.

THE ORIGINS OF THE MG HIVE OR IMPROVED NATIONAL HIVE

Recently I was asked why beekeeping examinations syllabi are somewhat vague as to the origins of the National and Modified National hives. My enquirer was all too aware that I had considerable contact over a period of years with two of our early pioneers of modern day manufacturing, namely Michael Burtt, grandson of the founder of Burtt & Son, Gloucester, and Arthur Abbott of Mountain Grey Apiaries (MGA), Brough, East Riding of Yorkshire.

Mountain Grey Apiaries and Arthur F Abbott

The historical information available regarding the origins of Mountain Grey Apiaries (MGA) is patchy to say the least.

Born on 6 August 1904, Arthur Abbott was the eldest of three sons of Mr and Mrs J Abbott, Station House, Brough, East Riding of Yorkshire. Abbot senior was a station master for the North Eastern Railway. At 13 years of age, young Abbott joined the Great Northern Railway in Doncaster. This was made possible through a family connection, his great uncle, JE Abbott, OBE, Trains Assistant to Superintendent of the Line, King's Cross. With the regrouping of the railways in 1923, the Great Northern Railway and the North Eastern Railway combined to form the London and North Eastern Railway (LNER).

Following the end of the Great War, railway companies suffered falling revenues because of competition from road transport, exacerbated by servicemen being given the opportunity to purchase surplus War Department open-top lorries when discharged from service. This massive increase in road transport created major problems for the railway industry which was a 'common carrier' and had to transport low value goods, whereas the new road transport industry could be selective and carry profitable high-value goods. Work in the railway industry in the early 1920s was hard to come by; family members were the first choice to fill any vacancies. In 1923, Abbott transferred to the Hull fish dock offices as a senior clerk.

From an early age, Arthur Abbott, like his father, was a keen beekeeper, his main hobby alongside playing tennis. He was a wealthy young man and gained growing respect in the local community. From 1924, he was the youngest delegate to the now defunct Hull and East Riding Beekeepers' Association. It was at one of the association meetings that he met the late Colin Wadsworth of 'Ashfield', Goole, a much-respected beekeeper and a prosperous businessman with ship building and brokerage interests. Arthur was an active member of the association and served on the committee. He was also a steward of the honey section at Brough horse show.

Arthur met his future wife, Marie Middlebrook, a trained elocutionist and the daughter of another prosperous Goole ship builder, at a summer tennis party at Mr Wadsworth's palatial riverside mansion, which had its own mooring on the River Ouse. Arthur and Marie were married on 12 June 1929 at St Mary's Church, Elloughton, East Riding of Yorkshire. They had a son, David, who is believed to have become a physicist at Windscale, Cumbria, and a daughter, Berris, who is still alive (2016) in nearby Leven.

Before 1923, Arthur had begun to expand his beekeeping activities into a prosperous commercial venture, helped by a loan from a close relative to purchase a small army surplus lorry in order to transport his bees to the clover which was abundant in the Yorkshire and Lincolnshire Wolds. The Wold areas were renowned for extensive sheep farming on many thousands of acres of Kentish white clover which provided abundant forage for bees. It should be noted that British farming was at a low ebb and livestock farming was more widespread than arable farming due to Empire preference tarrifs. When the weather was right it 'rained' honey. The base for the beekeeping business was Abbott's home, 'Rozel', in Brough. By 1936, his honey-producing business had become very profitable, so much so that he was able to send both of his children to well-known public schools in Yorkshire. In addition, he built new premises for the business at South Cave, Brough. This became his headquarters until ill health forced his retirement in 1961. The premises included honey processing facilities, workshops and storage, along with an office on the first floor of this commodious building.

It was through a meeting with an internationally known botanist, Revd M Yate Allen, in 1926, that Abbott became interested in Caucasian bees with their characteristic grey hairs on the tergal plates of the abdomen, prodigious hardiness and honey gathering potential. He obtained such

Mountain Grey Apiaries Limited

Honey Producers Association of Great Britain — Founder Member

BEES
HIVES
APPLIANCES

SOUTH CAVE — BROUGH, YORKS.

A Brief History, Folklore and Little-known Facts

bees for his own apiaries and queen rearing ventures. In May 1940, Revd M Yate Allen was interned on the Isle of Man for his involvement with far-right political groups. His associations with German and Russian beekeepers became a focus for suspicion under the infamous Regulation 18B for the state of the nation in wartime.

Earlier, in 1925, Abbott met Freddy A Wilkinson, an aircraft design engineer and accomplished draughtsman employed at the nearby Blackburn Aircraft Company. Wilkinson was also a hobbyist beekeeper and provided Arthur with detailed drawings for all the equipment that the company procured or manufactured over the following 30 years. Freddy, a fellow Freemason, joined the business full time when it became a private limited company, Mountain Grey Apiaries Limited, on 20 April 1944 with £2,000 nominal capital in one-pound shares and with Abbott as the controlling director.

With the boom in wartime beekeeping, there was a great demand for equipment. Fortunately, through his Freemason connections at Meyers Timber, kindled from his days in the railway industry, Arthur was able to obtain western red cedar timber without difficulty.

Following the end of European hostilities in May 1945, there was a change of government in July. Abbott informed the author that he had been told on good authority that the president of the Board of Trade, Harold Wilson (a future prime minister and known to be a pro Zionist), had been a consultant to Meyers Timber before becoming a member of parliament. The withdrawal of lend–lease aid from the

(Opposite) Cover of a Mountain Grey Apiaries product catalogue

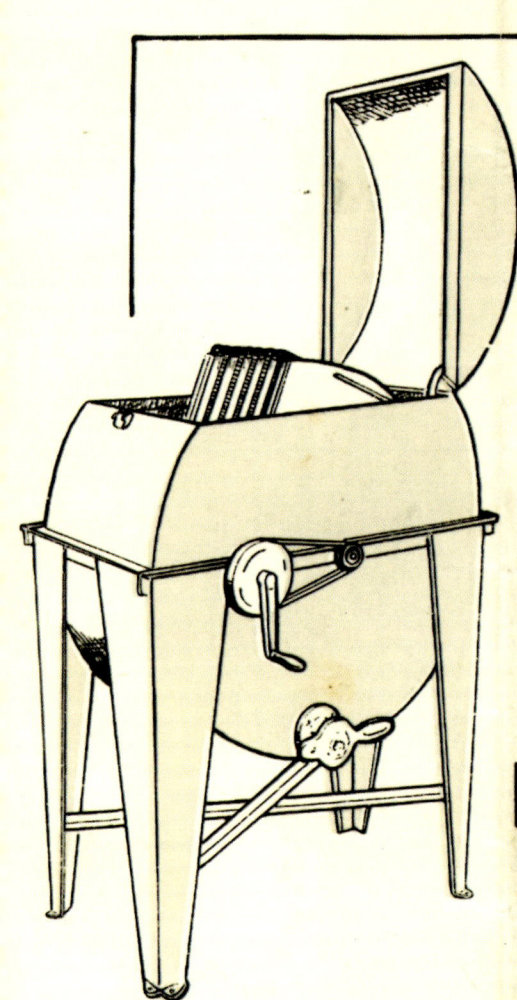

THE "MG"

PARALLEL - RADIAL

HONEY EXTRACTOR

The World's Speediest Honey Extractor

AN M·G INNOVATION

1948

A Brief History, Folklore and Little-known Facts

United States of America led to the troubled times of dollar importation which saw the introduction of licences for timber. No doubt this appointment and past loyalties gave Meyers Timber an advantage with regards to maintaining timber supply through its wharves at the major ports; a classic case of who you know, not what you know.

Abbott was an astute businessman who used bulk honey as currency. The late Cecil Tonsley told the author that Abbott made advance payment for advertisements in the *British Bee Journal* in 12.5 kg (28 lb) tins of honey. It was known that a cartel of the larger honey producers worked together to ensure that the price of honey was maintained and that there was no oversupply.

The 1947 Agriculture Act saw a revolution in farming techniques through research, science and subsidies. Incentives that included the application of high-nitrogen fertilisers led to the resurgence of ryegrass over low-growing clover and, by the 1950s, put an end to the large honey yields obtained from traditional Kentish white clover. This period coincided with a gradual decline in beekeeping that was to last for another 20 years.

From 1948, Abbott realised he needed to be ahead of the game. The woodworking machinery he was using was antiquated; it was inefficient and needed replacing. Being the hard-headed businessman that he was, he realised that he needed to reduce production costs in order to maintain a sufficient margin of profitability to stay in business. To this end he decided to outsource all of his timber work to

(Opposite) The 'MG' parallel–radial honey extractor as featured in a Mountain Grey Apiaries product catalogue

The 'MG' wax extractor and clarifier. This product was patented with British patent no 33194/45 and a Dutch patent no 128/252 [Michael Badger]

The 'MG' wax extractor and clarifier as featured in a Mountain Grey Apiaries product catalogue

THE "MG" WAX EXTRACTOR AND CLARIFIER

AN MG INNOVATION

1946

A Brief History, Folklore and Little-known Facts

woodworking and joinery specialist Bellwoods of Leicester. Burtt's chose a different route and kept manufacturing in-house; this was probably the cause of its demise. Since 1946, Abbott had used Triton Engineering, a small engineering company run by the Heaton brothers of North Cave, to manufacture the metalwork for MG honey presses, wax extractors and the three types of honey extractor he marketed worldwide. A big customer was the Crown Agents which bought both extractors and heather presses.

In 1954, Arthur took over the stocks of fellow commercial beekeeper Walter Allen, of Yorkshire Apiaries, Kirk Ella, Hull. In order to get round the timber licences introduced after the Second World War, Yorkshire Apiaries had diversified into portable prefabricated buildings. Allen, known to be a better businessman than commercial beekeeper, began to manufacture caravans as the Willerby Caravan and Touring Company in that same year.

In 1958, Abbott was becoming plagued with sciatica and angina. Unable to cope with the help of only one full-time employee, principal apiarist Ken Jackson, he chose to sell the business. Freddy Wilkinson had already retired. The business was sold to John (Eddy) Eade of Holme-on-Spalding-Moor. Jackson, who lived in Dundee Street, Hull, was asked to continue working for the new owner but declined owing to the long travel distance involved and the cost of public transport to Holme-on-Spalding-Moor. In addition, the pay was based on lowly agricultural wage rates. So, after 14 years with Mountain Grey Apiaries, Ken began a new career lasting 35 years with the Hull Telephone Service that gave him an immediate wage increase of five shillings (25 pence) a week and a saving on travel costs to boot.

Arthur Abbott (left) with Arthur Roantree (right) at an apiary at Land of Nod, South Cliffe Common, East Riding of Yorkshire, 1959 [William Slinger]

The loss of Ken saw Eddy Eade reduce his bee-breeding activities to concentrate on equipment sales and honey production.

The sale of the business to Eddy Eade did not include the limited company status. The original company eventually entered voluntary liquidation at the hands of the remaining members of the company, as reported in the *London Gazette*, 15 November 1966. By now MGA was a sole trading business. It ceased all beekeeping activities in the mid 1990s, but trading in bric-a-brac and house clearances continues under the name in the town of Goole, East Yorkshire.

Burtt & Son Limited

A number of years before Michael Burtt (third generation) died, he lived quite close to my parents in rural Warwickshire.

A Brief History, Folklore and Little-known Facts

This was convenient for me as the late Anthony Rawlins (a well-respected beekeeper and proponent of native bee breeding) resided in the same village; Anthony and I had similar interests in both bees and Alvis motor cars.

I recall Michael Burtt telling me about the history of the Burtt beekeeping empire from its earliest days. EJ Burtt & Son was known to exist as a beehive manufacturer in Stroud Road, Gloucester, from 1887. From 1894, it is also listed as a coal merchant with an office in Stroud Road and a depot at the Midland Wharf, where a number of rail wagons for coal carried the sign 'Burtt, Gloucester, Bee-hive Manufacturer'. Burtt & Son was mentioned in Railway Clearing House lists for 1926 but not after 1933. By 1935, the company was listed in *Kelly's Directory* as bee appliance manufacturers and picture frame manufacturers. The ancestry of the Burtt family is complex. They had their origins in Kettering, Northamptonshire, before moving to Gloucester. It is highly likely this move was to be near to the family of Burtt's wife, which owned a sack hire business in Gloucester Docks.

Edward John Burtt was born in Kettering in 1860. As a young man he worked for his maternal grandfather's sack hire business. He became interested in bees at a time when modern beekeeping was in its infancy. Indeed, so strong was his interest that, in 1886, he founded a beehive manufacturing company that was to bear his name for almost a century. He timed this venture well as it became one of the four main beekeeping appliance manufacturers in Britain. The business covered the west country, while other firms, Robert Lee in Uxbridge, EH Taylor in Welwyn, EH Thorne in Wragby and Steele & Brodie in Dundee, supplied the south, east and north of England and Scotland, respectively. Very soon, Burtt was to become one of Britain's best-known beekeepers. He was a founder of the Gloucestershire Beekeepers' Association, rising to the rank of honorary vice president at the time of his death in 1938. He also became a Victorian school reformer (no doubt because of his Quaker origins). 'EJ' was a champion of the adult school movement and the temperance movement.

As a major beekeeping appliance manufacturer, the seasonal nature of the business forced the company to diversify from the outset into coal and coke supply, which kept the income steady for the business, particularly during the winter months. Burtt & Son extended its carpentry skills to picture framing, and making furniture and Sunday school equipment. A catalogue surviving from 1933, printed by Burtt Brothers of Hull like much of the company's other literature, shows chairs for superintendents, adults, seniors and children as well as cupboards, collecting boxes, chalk boards, sand trays and models of buildings from the Holy Land.

Following the huge colony losses from 1904 to the 1920s, Burtt & Son was keen to produce a new hive that would appeal to the majority of beekeepers because of its reasonable cost. The company persisted with the concept of the National hive. All through the inter-war period, 1918–1939, the fortunes of Burtt & Son ebbed and flowed.

A beekeeping renaissance occurred in wartime in the 1940s but, sadly, by the 1950s fortunes were changing. The demand for honey dropped dramatically when sugar came off ration and changes in agricultural production practices also had an impact. The idea developed that, rather than making everything in-house, components made by outside firms for subsequent construction using a screwdriver might be the

A Brief History, Folklore and Little-known Facts

solution. Speculative steps into poultry-keeping appliances were also made at this time although this earned little income and was soon abandoned. A run of poor summers in the 1960s coincided with the rise of the do-it-yourself (DIY) movement, resulting in beekeepers building their own hives.

The late 1950s and early 1960s were depressing times for beekeeping. Burtt & Son ceased trading in Gloucester in 1974. Despite the cessation of the main business, Michael Burtt kept a limited trade in beekeeping appliances until 1986. The remains of the Gloucester operation were transferred to EH Thorne of Wragby, Lincolnshire, less than 80 miles from where the Burtts originated.

The National Hive

The concept of the single-walled National hive has its origin in the Congested Districts Board (CDB) hive produced during the nineteenth century. It was known that either side of the Great War, the Burtt family made overtures to the Ministry of Agriculture adviser, William Herrod-Hempsall, through its own large commercial bee farming enterprise. The family called for the introduction of a new single-walled hive, along the lines of American single-walled hives. Herrod-Hempsall was known to have a prejudice against anything non-British or not originating from the Empire. This prejudice was well known and is the reason why he was constantly in dispute with others. This eventually led to the break up of the British Beekeepers' Association in 1943.

The introduction of the government-sponsored 'bee restocking scheme' of 1919–1920 to overcome the losses that had arisen through the Isle of Wight disease epidemic and for other reasons (losses exacerbated by the absence of imports during the Great War), saw pressure mounting for an alternative type of hive. Herrod-Hempsall was resistant to the use of any hive other than the WBC double-walled hive. The Burtt family was not so easily deterred or, for that matter, toppled in its quest for the new 'National' type to be introduced. The Burtts were Quakers and above any forms of inducement to leave matters as they were with continued promotion of the WBC hive. Pressure from members of parliament saw a complete review of bee husbandry, including hive types and importation of bees, as a means of stabilising colony losses.

The introduction of the new National hive took off when it was agreed that it was two-thirds cheaper to manufacture and cheaper to maintain than the WBC hive. Reported issues of dampness and poor ventilation were found to be spurious and, above all, transportation was so much easier. Herrod-Hempsall conceded that the new National hive did have an air space at each of the two ends where the frame lugs met. Despite substantial evidence from the United States of America and Canada, Herrod-Hempsall would not acknowledge that such hives fared well in cold climates and that colonies in those hives that benefited from warming sun in the spring were stronger than those housed in double-walled hives.

Within a decade, the new National hive was to be supplanted by the 'MG' Modified National hive.

The 'MG' Hive or Improved National Hive

While skimming through ROB Manley's book, *Beekeeping in*

A Brief History, Folklore and Little-known Facts

Britain, I came across a piece about the origins of the National hive and the Improved National (pages 122 to 123):

> *Lately a very sensible hive of this type has been made by Mr AF Abbott. It is a new form of what has been sold as the 'Improved National'. Personally I am attracted by the hive made by Mr Abbott.*

In 1983, the late Colin Weightman and I visited the late Arthur Abbott at his home in Leven, East Yorkshire, to discuss the various historical points of the parallel–radial honey extractor. (Abbott never patented this. It has since been developed and improved by Maxant of the USA.) He had perfected this extractor in 1946, based on the BOMNS honey extractor of the USA. In the course of the discussion, he mentioned the origins of the Modified National hive, remarking that few people ever asked about or questioned its origins.

It transpired that he had employed several part-time specialist woodworker machinists for making beekeeping equipment. Interestingly, in 1925–1926, these workmen originally manufactured sea plane parts at the Blackburn Aircraft Company (now British Aerospace) opposite Whitton Ness, at nearby Brough, for the ill-fated 'Blackburn Iris', a three-engined three-bay biplane flying boat of which only five were manufactured. These artisans came from the furniture trade and were highly skilled, used to working with precision to high standards. Most of them lived in High Wycombe, Buckinghamshire. With work at Blackburn's ceasing at midday on Saturday until Monday morning, these workers, far from home, were 'kicking their heels' all through the weekend. Abbott realised these skilled workmen could turn out all the gear needed on a Sunday, freeing up his own men to concentrate on honey production during the working week. Arthur mentioned that one of the machinists suggested a revision to the National hive to him, which was perfected into the now Modified version. The machinist was not a beekeeper but thought that the machined hand holds of the National hive were hideous. Freddy Wilkinson drew up a perfected cutting plan. It was not until 1927 that Arthur realised this hive's potential following a visit to his works by a very large commercial beekeeper from Norfolk who saw this prototype hive lying on a workbench gathering dust and enquired as to its origins. Taken in by its simplicity, its robustness and the ⅞ in (23 mm) thickness of the walls, an order was placed for 1,000 deep boxes. The new Modified National hive gained immediate popularity.

The outbreak of the Second World War in 1939 saw a terrific upswing in demand for beehives. Abbott's work in pollination resulted in him being granted 'reserved occupation' status, enabling him to continue with his beekeeping enterprise.

Hull and its docks, like Plymouth and Liverpool, became a target for heavy bombing which required a suitable workforce to make damaged homes habitable again. Abbott's workmen, with their woodworking skills, saw the opportunity to earn better pay by becoming 'war damage repair artisans'. To make up for the loss of these workmen, Abbott enlisted the services of young evacuees to assist him with beekeeping tasks; farmers were known to be present at evacuee dispersal points in order to pick out the strongest looking young men for manual farm work.

Beekeeping was booming, assisted by the introduction of sugar rationing. With many beekeeping clubs increasing their membership numbers, demand for Abbott's Modified National hive outstripped his labour resources. The solution came from an unexpected source. Arthur Abbott had forged a close relationship with another Yorkshire beekeeper, George Green, a fellow Freemason and also the 'Green Eye' correspondent for the now defunct *British Bee Journal*. At the time they were engaged (with others) in trying to bring together the Yorkshire County Beekeepers' Association and Yorkshire Beekeepers' Association as a unified body. This relationship was mutually beneficial in enabling the demand for hives and equipment to be met. At the time, George Green was the general manager of Harehills School for the Disabled in Leeds. He was able to put his workforce to good use which, in turn, was to see this demand for beekeeping equipment met by giving those with disabilites meaningful and enjoyable work.

Despite wartime timber shortages, fate took a hand once more with the close proximity of the ports of Goole and Kingston-upon-Hull. Timber of all types, grades and quantities was landed at the wharfs of Meyers Timber, the largest timber importers in Great Britain. Abbott was able to collect timber from the docks and deliver it to the workshops in Leeds. On the return journey, beekeeping gear prepared 'in the flat' would be carried back to South Cave for assembly. Transporting the unfinished items enabled Abbott to carry 30 times the amount of already assembled equipment.

The success of this enterprise saw George Green awarded an MBE in 1945 for his good works for the disabled. Arthur said with a chuckle, '... and I did not do too badly out of it too; I bought a villa close to St Antoni de Portmany, Ibiza'.

THE "MG" AND BRITISH STANDARD NATIONAL HIVE

AN M·G INNOVATION

1927

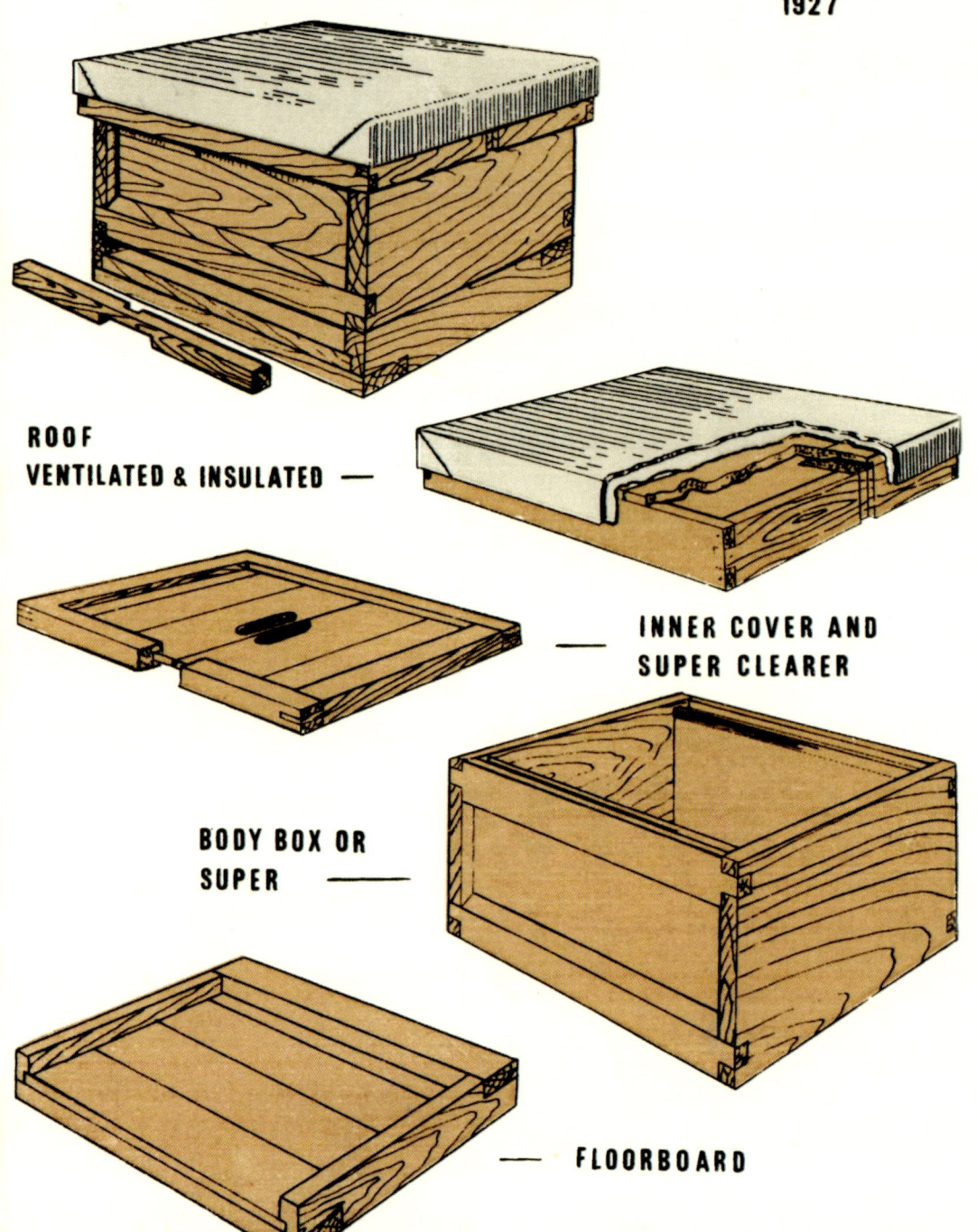

ROOF VENTILATED & INSULATED

INNER COVER AND SUPER CLEARER

BODY BOX OR SUPER

FLOORBOARD

A Brief History, Folklore and Little-known Facts

Arthur Abbott did not seek a patent on the Modified National hive, as he was advised to do. He was content to collaborate with the British Standards Institution (BSI), along with the various beekeeping organisations. He gave the BSI the rights to become the registered owner of a certification trade mark. Beehive manufacturers could use this, under licence, on hives which complied with the applicable British Standard, BS 1300:1946, with a succession of amendments up to 1960.

Mountain Grey Apiaries continued to manufacture the hive in 25 mm (1 in) Canadian western red cedar finished to 23 mm (⅞ in) thickness. However, the British Standard specified the use of 19 mm (¾ in) thickness and this was adopted by most of their competitors.

The introduction of the British Standard for this hive saw it renamed as the 'British Standard National Hive', although the term 'Modified National' is still used as the original National hive can still be found, although in relatively low numbers.

WORLD WAR II: A REVIVAL IN BEEKEEPING

The outbreak of the Second World War saw the introduction of the food rationing orders which required, amongst other things, controls over the provision of sugar. These imposts resulted in a great influx of new beekeepers, which took membership numbers of local beekeeping associations to unprecedented levels. Those dark days saw many take to self-help in the Dig for Victory campaign. Everyone was

(Opposite) The 'MG' and British Standard National hive as featured in a Mountain Grey Apiaries product catalogue

encouraged to work together. Lord Woolton, Minister of Food, made weekly presentations on the theme of reducing waste and being self sufficient. With a background in the retail trade, he had a natural way of knowing what the housewife of the time wanted. In his endeavours to get the public at large to undertake a frugal existence, a pie was named after him: 'Woolton Pie'. This was an adaptable vegetarian dish, created at the London Savoy Hotel by its then Maître Chef de Cuisine, Francis Latry. The bill of fare for the day at Simpson's-in-the-Strand, 8 May 1945, was 5/- (25 decimal pence). In these days of health consciousness, it is a wonder that the Woolton Pie has not made a comeback!

Special Sugar Feed Arrangements for Beekeepers during World War II

The following is an abridged extract from the article 'Friends in High Places' by Erica Osborn (*Bee Craft*, 87(3), March 2005).

> *In his memoirs, Sir Winston Churchill mentions that on 9 April 1943 he sent an edict to the Minister of Food, Lord Woolton: 'What is the amount of sugar issued to professional beekeepers, and what is the saving in starving the bees of private owners?'*
>
> *Why did this come about? Why was the Prime Minister concerned with bees and beekeeping when, at the time, the Battle of Tunisia was in full cry?*
>
> *A well-known beekeeping personality at the time, the late Austin Hyde, who was then the headmaster of Pickering Grammar School, was an enthusiastic beekeeper and renowned broadcaster on the North Riding*

A Brief History, Folklore and Little-known Facts

Daphne Birch, Ripon, Dennis Robinson, Shipton-by-Beningbrough, and Austin Hyde, Pickering, at the National Honey Show, 1959 [Bernard Leafe]

dialect. Hyde was no fool when it came to beekeeping; the introduction of food rationing was to have wide implications, not only for man but for livestock too. A reduction in the sugar ration meant certain death to his bees, due to the lack of forage in his part the country (the Vale of Pickering). It must be realised that in wartime conditions all pasture was put to the plough by the War Agricultural Executive Committee (affectionately known as the 'War Ag'), hence there was very little clover which is a useful source of nectar for the bees' winter stores. The reliance on ling heather for overwintering is, to say the least, very chancy. With this in mind, Hyde appealed to the highest quarter.

The crie de coeur from the North Yorkshire moors managed to reach the prime minister, Sir Winston Churchill,

Food order certificate for obtaining sugar for feeding honey bees (front)

Food order certificate for obtaining sugar for feeding honey bees (reverse)

A Brief History, Folklore and Little-known Facts

himself. His memoirs clearly show his indignation at the possible loss of his favourite breakfast food and he issued the necessary directive that sugar be made available to beekeepers.

So who was the mystery person to whom Hyde appealed? It was none other than the Rt Hon William Mabane who, in 1943, as parliamentary secretary to the Ministry of Food, visited Pickering to open a War Savings Week. Mabane had been Hyde's pupil at Woodhouse Grove School, Apperley Bridge, Bradford, many years before.

When visiting London for the National Honey Show during November of the Coronation year, Austin Hyde called at 10 Downing Street to leave a taste of Yorkshire honey 'with the gratitude of the bees'. A letter was subsequently received expressing warm appreciation.

Trips to the Moors in Wartime

Despite petrol rationing, the ingenuity of a number of well-placed beekeepers enabled them to arrange transport for their bees to and from the moors. A good number of beekeepers, who were privileged to have access to the moors, banded together through beekeepers' associations to take their bees to the ling with the object of surplus honey and to get the bees fed at nature's expense, with the beekeeper's wife getting the sugar for household consumption. When sugar came off ration, it led to a sharp decline in the number of beekeepers, with some beekeepers' associations being disbanded by the early 1950s. Nowadays, personal vehicle ownership gives beekeepers the freedom to make their own arrangements for taking bees to the moors. The reliability of

Heather Honey: A Comprehensive Guide

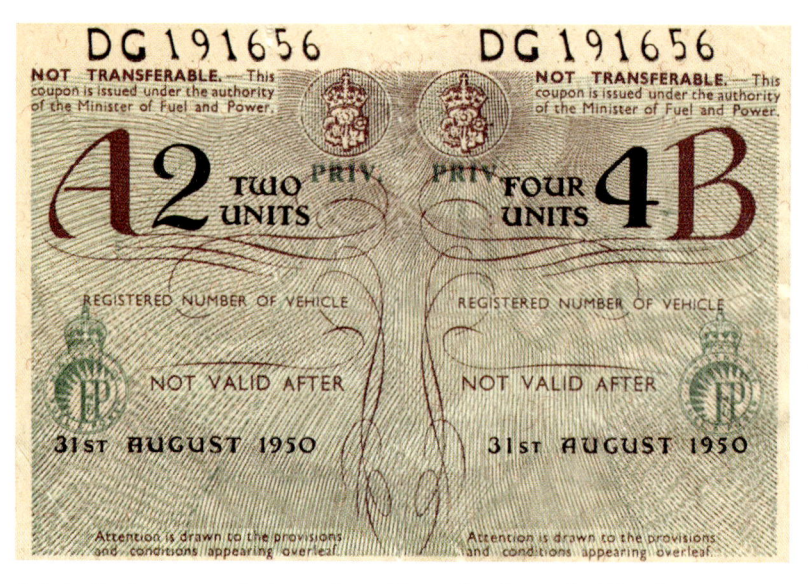

Petrol rationing coupons for two units and four units of motor fuel (front)

Petrol rationing coupons for two units and four units of motor fuel (reverse)

vehicles available today makes such journeys less demanding; not like they were, not so long ago.

The War Years: Austerity and Making Do

Most beekeepers' associations produced a printed news-sheet of some description that was circulated by volunteers or by post. An issue of a newsletter was a welcome arrival.

The wartime spirit stirred many to bring humour into the lives of others. Rowland Wood was a desk sergeant at a Leeds police station. Through his daily contact with the public, he derived an impish sense of humour that he conveyed through his writings on beekeeping. This kindly man took a lighthearted look at matters with his writings that began in 1942 under the *nom de plume* of 'Lignum'. He published a series of humorous ditties entitled *The Chronicles of Lignum*. The following is a parody about his beekeeping colleagues on their trip to the heather moors, published in the Leeds Beekeepers' Association house magazine, October 1944. This particular article relates to going to the heather moors at Pateley Bridge, North Yorkshire.

The Chronicles of Lignum – Chapter VI

And on the appointed day, the Onsec called forth unto the strong men that they journey unto the mountains of Patelei whereon grew the heather.

And there came forth many men of mighty girth.

And the Onsec exhorted them and said unto them: Though to-day be the Sabbath, beget thou thine oldest raiment and wear of it. Also carry thou food in plenty, for the journey is long and the tasks ahead of thee hard.

And they answered and said unto him: Though our wives upbraid us, yet shall our raiment be nigh unto sackcloth.

And it was so.

And ere the sun was risen, and while it was still dark, these mighty men came forth from their tents and set forth unto the mountains of Patelei.

And they took with them many hives of bees, painted, un-painted and newly-painted, from many vineyards.

And each of them boasted of the rags that he wore. Whereupon Walter, of the tribe of Butterworth, of sugar fame, lifted his Civil Defence overcoat and cried unto them: Behold ye the pants of my grandfather and boast ye not of your feeble rags.

And they hung their heads in shame for never had men seen the like before.

And when they came unto the mountains whereon grew Calluna vulgaris, beloved of the bees, they toiled mightily and were weary. And some among them went forth to gather blackberries but others cried: Behold, here is nectar for the bees. Let us go forth in search of nectar of man.

And they searched diligently and came unto the vineyard of Hammond, wherein was much nectar from the hop, and all were comforted.

And they set forth from the mountains, all of them, even the pants of Butterworth, rejoicing.

Consider ye these hives from the vineyards, what manner of array is this.

Noble pagodas to charm the eye; converted dog kennels; family chests; nailed hen coops; small garages; large soap

A Brief History, Folklore and Little-known Facts

boxes; cabin trunks; builders' huts; cardboard cartons; enemy search points; linoleum flight decks; hives with concrete portals; hives secured by the rocks of Brimham; tall hives of hope; squat hives of despair; mansions; villas; flats and hovels.

But which of you, reading these words shall mock such display? For have not the wise ones many times been asked the riddle: Lovest thou the National or the W.B.C.? And is it not true that the wise ones have quarrelled nigh unto bloodshed over the riddle?

And who shall cry shame upon the lowly beekeeper that converteth his coal box into a hive until the riddle be solved.

W.R. Wood

HEATHER HONEY FOR MEDICINAL PURPOSES

The narrative below is written in layman's terms and is based on the Author's own understanding of honey for medicinal purposes.

Honey is known to possess three biological properties that make it unique with respect to its antimicrobial properties. They are:

- a glucose-oxidase system. During the process of honey production, worker bees add the enzyme glucose oxidase to the nectar. When honey is applied to the wound, this enzyme comes into contact with oxygen in the air, which leads to the production of the bactericide hydrogen peroxide
- honey is a hyperosmotic agent that draws fluid from the wound bed and underlying circulation. This kills bacteria that cannot thrive in such an environment
- honey is an acid medium; it is predominantly gluconic acid. The pH range of honeys from different sources and of different types (floral, tree, or ling heather) is 3.1–4.5; the average is regarded as 3.9. Acidity within this range precludes the establishment and growth of many species of bacteria. Gluconic acid is an organic compound derived from glucose by the enzyme glucose oxidase.

To be effective the honey should contain a minimum of 20 per cent moisture.

Heather Honey for Medicinal Purposes

The use of honey as a medicinal aid can be traced back to Egyptian and Roman times. Honey was used to treat infected wounds as long ago as 2,000 years ago, before bacteria were discovered to be the cause of infection. Pedanius Dioscorides (*circa* AD 40–90) was a Roman physician, pharmacologist and botanist of Greek origin. He was the author of *De Materia Medica*, a five-volume encyclopaedia of herbal medicine and related medicinal substances and the forerunner of modern day pharmacopoeia. This was probably the first magisterial work on the subject. It was known to be read far and wide for more than 1,500 years. Dioscorides describes honey as being 'good for all rotten and hollow ulcers'.

The appreciation of the value of honey as a wound dressing goes back to the ancient physicians whose modern medicinal knowledge of germ theory was wanting. They came to realise that honey was antiseptic and antibiotic. Honey today is widely used by both medics and veterinarians.

My father served with a cavalry regiment of the British Army, (1st) Royal Dragoons (now the Blues and Royals). Prior to the outbreak of the Second World War his regiment became mechanised. Being an experienced trooper and horseman, he transferred to the Royal Army Veterinary Corps (RAVC). He was a natural with animals. To his dying day he never drove a car or, for that matter, flew in an aircraft. My father was typical of his generation; as he grew old it was recollection of memories that kept him going. I recall my father citing many instances of veterinary officers who were advocates of using honey (to great effect) for serious wounds on mules and horses. He mentioned that one of the officers to whom he was attached, Major (later Brigadier General) Hector Wilkins, was a strong believer in honey as a medicament (due to its antimicrobial properties) to aid healing. Soon after the Second World War, Major Wilkins headed up the zoological section at Porton Down for research into the protection of animals against chemical warfare.

In another sphere, my great uncle commented to me that, in the First World War, he came across severely injured Russian soldiers from the Bulgarian front being treated with honey dressings for serious gangrenous wounds. The use of antibiotics was still to be researched; yet Russian field warfare medics were alive to the use of honey for its antimicrobial properties.

NEW HONEYS FOR USE IN MEDICATION

In the past 20 years, honey has seen a resurgence in its medicinal use and it is becoming favoured by adherents of modern day medicinal pharmacopoeia. Ling heather honey from an upland moorland environment is of particular interest because of the absence of pesticides and other contaminants associated with modern day farming techniques.

From a layman's perspective, it is a fair statement that honey has both historical and traditional uses for the treatment of infected wounds. There are dozens of research papers on the subject that illustrate that honey can be effective against many antibiotic-resistant strains of bacteria. These papers identify that honeys produced from many different floral sources demonstrate that antibacterial activity can, and does, vary with the origin of the honey and subsequent processing. Honeys selected for medical use would need to be evaluated on the basis of their antibacterial activity levels which can only be determined by specialist laboratory testing. The issues concerning methicillin-resistant *Staphylococcus aureus* (MRSA), a bacterium responsible for several difficult-to-treat infections in humans, give rise to many research papers endorsing honey's use in overcoming antibiotic-resistant bacteria.

Care has to be taken in the use of honey. Honey produced as a food is generally not well filtered and may contain various particles that are best removed before using honey medicinally. Although honey does not allow vegetative bacteria to survive, it does contain viable spores, including those of *Clostridia*. Honey that has been treated by gamma irradiation is available commercially; this process kills such spores, without loss of any of the antibacterial qualities of the honey. It is an absolute necessity that only sterile honey is used in all clinical applications.

A detailed scientific and medical discussion on the uses for honey in medicine is beyond the scope of this book. This narrative is given to illustrate that honey has had a place in medicine in the past that is likely to continue into the future as science and understanding develops.

THE PRACTICAL CONSIDERATIONS FOR THE CLINICAL USE OF HONEY

Medical advice should always be sought when using honey in a clinical context.

- The quantity of honey that is required for wound management and treatment has to be related to the amount of fluid that may be issuing from the wound. The term for the discharge is exudate; that is, a fluid

that has exuded out of a tissue or its capillaries due to injury or inflammation.
- The frequency of dressing application is directly related to how swiftly the honey is being diluted by the exudate.
- Dressing applications, as a rule, need to be changed twice weekly if there is no discharge as a means of maintaining a 'reservoir' of antibacterial components as they dissipate into the wound tissues.
- The best results are achieved by applying heather honey to an absorbent dressing as the honey will warm quickly before a secondary dressing is applied and secured in place.
- Heather honey will not be absorbed readily into dressings; it may require warming to body temperature.
- Depending upon the depth of the wound cavity, simple adhesive film dressings can be used, with a smear of honey on the dressing area. However, for a moderate to heavy discharge, a secondary dressing is the norm to prevent the diluted honey from seeping through the primary dressing.
- Occlusive dressings were introduced in the USA in the 1970s. These dressings have a moisture vapour permeable polyurethane film. They are both airtight and watertight.
- The problem of a heather honey dressing sticking to a wound can be overcome by using a low-adherent dressing that is placed between the wound and the heather honey dressing. These particular dressings must be porous to ensure the antibacterial

Heather Honey for Medicinal Purposes

components of the honey can dissipate freely into the bed of the wound.
- The use of alginate dressings, derived from seaweed, with heather honey is an excellent alternative to cotton/cellulose dressing material as the alginate is highly absorbent; it is also biodegradable.
- The wound bed may have depressions or cavities that can be readily filled with honey, in addition to the use of the heather honey-impregnated dressing, thereby ensuring the antibacterial components of the honey are diffused into the wound tissues.
- A final note. Since infection may also lie in the tissues underlying the wound area margins, the honey dressing will need to be extended beyond the inflamed area of the wound.

The antibacterial properties of honey include the release of low levels of hydrogen peroxide. Some honeys have an additional phytochemical antibacterial constituent. Heather honey is regarded by many authorities as one of the superior honeys for clinical purposes.

Honey is produced from many different floral sources. Its antibacterial activity varies with its origin and the method of processing. Dioscorides stated that a pale yellow honey from Attica was the best; Aristotle (384–322 BC), when discussing different honeys, referred to pale honey as being 'good as a salve for sore eyes and wounds'. Today, honey from the strawberry tree (*Arbutus unedo*), of Sardinia, is valued by specialists for its therapeutic properties. In India, honey from the lotus (*Nelumbo nucifera*) is said to be a panacea for eye diseases. Honey from the sidr tree (*Ziziphus spina-christi*), of the Jordan valley, known locally as Christ's Thorn Jujube, is revered in Dubai for its therapeutic properties. Manuka honey has a longstanding reputation in New Zealand. Heather honey, especially ling, must also rank highly with the aforementioned honeys.

The principal benefits of honey, now included in some proprietary wound dressings, mainly results from the high sugar concentration dehydrating the bacteria by osmosis, poisoning them by acidity and, less certainly, through oxidation. Many authorities on the subject believe that honey's effectiveness is really nothing to do with the type of honey that is used, whatever certain honey producers would have you believe. It is a function of the sugar content. It is doubtful if there are any studies that have compared honeys head-to-head for effectiveness.

In these days of resistant bacteria, a significant problem in longstanding infections, honey is a treatment that gets round this problem.

Over the past century, considerable research has been undertaken regarding desensitisation to pollen. The small amounts of pollen in honey cause an IgE reaction (Immunoglobulin E [IgE] is a type of antibody). It is present in minute amounts in the environment and triggers the release of substances from mast cells that can cause inflammation in hay fever sufferers. Repeated ingestion of honey overwhelms the IgE response that leads to the allergy. For this reason, it is most effective with honey from locally growing plants that are the cause of the hay fever.

APPENDICES AND INDEXES

Honey bee working ling heather flowers [Brian Nellist]

APPENDIX A: FERMENTATION AND HYDROXYMETHYLFURFURAL (HMF)

Because of its naturally high water content, heather honey will spoil readily through fermentation. Mention has been made in the text of how the beekeeper can prevent or reduce the likelihood of this happening. Fermentation is a result of the conversion of the sugars in honey, to carbon dioxide and alcohols, under favourable conditions. These are a warm temperature and excess water content within the honey.

If honey of water content less than 20 per cent is kept at an ambient temperature below 10 °C (50 °F), the likelihood of fermentation is minimal or non-existent.

ENZYMES IN NECTAR AND ITS CONVERSION TO HONEY

The honey bee adds a number of enzymes to nectar while it is contained within her honey sac. It is beyond the scope of this book to explain the mechanisms involved in detail, other than to say that enzymes play an essential role in the conversion of nectar to honey and also in its functional properties. The enzymes of honey make it truly unique. Honey is far more complex than most, if not all, sweeteners.

Enzymes are complex proteins that are formed within living cells. They have many attributes that are involved in processes and reactions within living materials. Honey contains about a dozen unique enzymes in small amounts. These are listed in the table.

Enzymes of honey that vary widely by floral source

Enzyme	The chemical reactions catalysed
Diastase	converts starch to dextrins and maltose
Amylase	transforms starch into other carbohydrates such as oligosaccharides, disaccharides, monosaccharides and dextrins. It is found in nectar and is added by the honey bee during foraging and in ripening of nectar
Invertase	catalyses the hydrolysis of sucrose to glucose and fructose (invert sugar)
Sucrose hydrolase	catalyses starch and sucrose metabolism
Glucose oxidase	catalyses the oxidation of glucose to hydrogen peroxide and D-glucono-o-lactone
Catalase	catalyses the decomposition of hydrogen peroxide to water and oxygen
Acid phosphatase	frees attached phosphate groups from other molecules
Protease	hydrolyses proteins and polypeptides to yield peptides of lower molecular weight
Esterase	splits esters into an acid and an alcohol by hydrolysis
ß-glucosidase	catalyses the hydrolysis of beta-D-glucosides and oligosaccharides to release glucose

The level of activity of enzymes such as diastase has been used for many years as a means to ascertain the extent to which honey has been heated. More sophisticated methods which measure levels of another indicator, hyrdoxymethylfurfural (HMF), are now used.

Appendix A: Fermentation and Hydroxymethyfurfural

THE PROCESS OF FERMENTATION

The process of fermentation comes about through the conversion of the sugars by the yeasts present within the honey. These yeasts are naturally present in the nectaries of flowers; they are picked up and carried by the bee in the nectar. The yeasts grow and multiply using the sugars as the required source of energy. Growth and multiplication of the yeasts produce many by-products which impart flavours to the honey that blight its aroma and its natural flavour. It is beyond the scope of this book to explain the mechanisms involved in detail, other than to say that many of these yeasts die when the conversion from nectar to honey takes place because of the rise in concentration of the sugars, although some yeasts do survive. Their presence in the right conditions allows them to multiply to a point where their existence is a destructive force to the honey.

Yeasts are inhibited from developing at a temperature of below 10 °C (50 °F) and above 27 °C (80 °F). Therefore, fermentation can be kept at bay with confidence whether honey is stored in bulk (honey stores better in bulk) or in jars, under these conditions.

A water content below 20 per cent minimises the risks of fermentation. The concentration of the sugars is such that yeast growth and multiplication is inhibited. Honeys that become crystallised retain fluid between the crystals with an increased water content in the range 4–6 per cent. Therefore, such honeys can readily ferment if the ambient temperature is high enough. There are honeys in which fermentation can be held in check by their texture, especially hard crystallised honeys, whereas fermentation will be more likely in soft-textured honeys.

Fermenting honey has a distinct smell reminiscent of decaying apples; the odour is never forgotten by beekeepers who will know when their honey is on the turn (fermenting). Fermentation occurs in crystallised honey for three reasons. The first is a non-airtight container that permits water vapour to be absorbed by the honey from the atmosphere due to its hygroscopic nature. The process leaves a very thin layer of dilute solution on the surface of the honey that is the ideal medium in which fermentation can proceed. This watery layer can be removed; it has an unmistakable wine-type odour. The remaining unfermented honey can be processed or used as normal. Care is needed to ensure that all the fermenting material is removed and that the remaining honey does not become contaminated.

A second cause of fermentation is warming honey for processing. The fermentation is not obvious until the honey is poured into the bottling tank. Large bubbles are seen with the unmistakable smell of fermentation becoming apparent. The beekeeper can retain this honey by heating it to 95 °C (200 °F) and then it can be fed back to the bees for them to reconstitute. Under no circumstances should it be offered for sale.

The third cause of fermentation is thought to arise from the differing amounts of fructose present within the honey. The honey surface heaves like baker's dough, despite its fairly dry appearance. The smell of fermentation gives it away, together with the lumpy appearance. Careful removal of the top 15 mm of the affected surface will be sufficient to enable the remaining honey to be processed.

Appendix A: Fermentation and Hydroxymethyfurfural

HYDROXYMETHYLFURFURAL (HMF)

Hydroxymethylfurfural (HMF) is produced by the chemical breakdown of fructose in the presence of free acids. This occurs in honey all the time. The rate of HMF production is largely dependent upon the temperature to which honey is subjected at any given time. The higher the temperature, the faster HMF will be produced. The longer honey is stored before it is used, the greater the quantity of HMF it will contain. The amount produced naturally is relatively insignificant and it is completely harmless to the consumer. Maximum allowable levels of HMF are specified in the honey regulations. The USA does not have specific honey legislation and sets no legal limits for HMF or diastase activity, the quality being guaranteed by importers.

   Beekeepers may be ignorant of HMF and its presence in bulk stored honey. However, with normal and careful handling, the amount of HMF will not exceed the permitted level. Trading Standards officials are known to check honey available at retail outlets. Testing for the level of HMF to the necessary level of accuracy requires sophisticated laboratory processes. Therefore, it is really beyond the resource of most beekeepers to test for this. It is suggested that the beekeeper treats honey in such a manner as to reduce the risk of a significant increase in HMF level. A simple solution is to ensure that bulk stored honey is processed in date order (oldest first), that it is stored in cool conditions and that any heat treatment is minimised.

APPENDIX B:
THE DYCE METHOD OF PRODUCING CRYSTALLISED HONEY

The production of naturally crystallised honey is somewhat of an anachronism when the production of beautifully prepared soft-set honeys is in vogue. However, although diminishing, there remains a market for such honey.

Many beekeepers fall well short of producing a crystallised honey of merit. The large honey shows feature such honeys with few samples being awarded prizes, due, in part, to ignorance of how to prepare the perfect sample.

In the early 1960s, I was present at a lecture given on the subject by a colleague of the late Professor Elton James Dyce from Cornell University, Ithaca, New York. Professor Dyce died in 1976. At the time, he was regarded as the foremost authority in the world on the commercial preparation of honey for the retail trade. My notes of the lecture were written up immediately afterwards.

The processes and plant layouts of the large commercial honey processors are, in the main, based on his concepts for production of crystallised honey.

Professor Dyce commenced his beekeeping as a student at the Ontario College of Agriculture at Guelph, Ontario, Canada, later graduating and undertaking further research as a postgraduate student in beekeeping. At some stage, he joined the teaching staff. Some time later, he moved to Cornell University as an assistant to Dr EF Phillips, whom he was to succeed. Professor Dyce is credited with and best known for the discovery of the precise circumstances under which a

Appendix B: The Dyce Method

commercially acceptable crystallised honey could be produced in a standardised form. This method is now the accepted method in general use by commercial honey producers worldwide. The Dyce process overcomes the issues of partial granulation, fermentation, excessively coarse granules and the hardness associated with naturally granulated honey.

The process is described below.

- The honey is heated to a temperature of 49 °C (120 °F) to dissolve any natural crystals by putting the bulk containers into a temperature-controlled warming cabinet.
- Once the honey has liquefied, the temperature of the honey is allowed to fall below 27 °C (81 °F).
- An amount of the finest grained crystallised honey is added as 'seed' at the propotion of 500 g of seed to 13 kg of the warmed bulk honey. The seed honey needs to be warmed first so that it will mix readily and the whole mixture is stirred thoroughly to ensure that the seed is fully incorporated. It is essential that the seeding is undertaken when the bulk honey is at a maximum of 27 °C (81 °F). A higher temperature will dissolve the crystals of the seed, thereby aborting the process.
- The bulk honey is allowed to granulate to solid at a temperature of 10–14 °C (50–57 °F). A consistent temperature is a critical factor for success. The use of a well-insulated cabinet is essential. The process may take up to ten days. The set honey will be smooth, provided that the crystals in the seed were fine. Once the bulk has solidified, it may be found to be too

hard. To overcome this a softening process will be necessary.
- The set honey is warmed again to a maximum temperature of about 30 °C (86 °F) within the insulated warming cabinet. Once it begins to soften, the honey should be stirred thoroughly. Manual honey-creaming devices can be used but it is essential that no air is incorporated into the bulk honey by paddling the honey and lifting the creamer too far from the surface. This will tend to draw air into the honey. An improvised paddle can be constructed that fits into the chuck of a controlled electric speed drill. The makeshift paddle is made of a piece of 20 mm x 3 mm bright steel bar secured to a spindle of 6 mm diameter bright steel rod.
- The softened honey should be stood in the warming cabinet for several hours at a temperature of about 25 °C (77 °F) to allow any incorporated air to rise to the surface. It is then bottled. Care must be taken to prevent overheating which will cause the honey to reliquefy. Individual set ups may vary so some practice is advisable. Carried out correctly, the jars of honey will be free from 'frosting'.

APPENDIX C:
A METHOD OF PRODUCING SOFT-SET HONEY WITH LING HEATHER HONEY

As a honey judge of almost 40 years, I am asked to judge at many honey shows throughout the United Kingdom. On my travels, it never ceases to amaze me that the majority of beekeepers have no real idea of how to produce soft-set honey, whether for the retail trade or for exhibition purposes.

The production of soft-set honey is a relatively simple procedure. The process is anything but difficult to achieve provided that the principles involved are understood. It is a process in which the crystals within the mass are broken down by pulverising or mashing, leading to the honey attaining a spreadable consistency that mirrors soft margarine. It has to be understood that this spreadable consistency cannot be achieved through reliance on natural crystallisation.

A honey that is in a set or solid state that has been liquefied by heating and then is allowed to reset invariably granulates with coarse asymmetrical crystals that are unusually large. Seeding, as discussed in Appendix B, is an accepted process that causes honey to granulate with a fine crystal size, resulting in an acceptable smooth texture. Proficiency comes through practice and using a good seed honey. All beekeepers can become excellent producers of soft-set honeys in a relatively short time.

It is generally accepted that naturally granulated honey is not customer friendly. It is really a product confined to the show bench or exhibition. Its purpose these days is to demonstrate the skill of the exhibitor. It has to be said that very often its appearance is anything but attractive, especially if the honeys used are too dark. Clover honey is the best option. Other physical features that detract from its appearance are:

- upon the honey being run into the jar, it shrinks away from the jar wall when it sets, giving a 'frosted' appearance that may be quite extensive
- a coarse and unbalanced product may be produced by uncontrolled natural granulation which often results in a product that is far from the fine and smooth uniform texture desired
- a hard texture in the jar that requires excessive force to remove it that can cause the spoon to bend
- a honey that is often only partially set and apt to ferment more readily (see Appendix A).

The production of soft-set honey is not readily achieved with high fructose honeys. To ascertain how the honey will granulate, it should be processed by extracting, filtering and storing in 15 kg food grade plastic containers, complete with sealed snap-on lids that are completely airtight. The containers are then properly labelled and dated. The honey is stored in a cool place, out of direct sunlight and, after three months, its condition is checked, particularly its granulation.

A word of caution about plastic buckets that are filled with 15 kg of solid honey. When handling full buckets of honey, treat them with care and avoid rough handling such as dropping the containers onto solid floors. Damage to the container may not become apparent until the honey is heated.

Appendix C: A Method of Producing Soft-set Honey

Those containers that have only partly granulated should be clearly identified, labelled and dated, and set aside for processing into clear honey. Honey in stored containers that has granulated needs to be identified as to the quality of the grain, whether fine and smooth or having large coarse crystals.

There may be situations when the top of the bulk stored honey has a 'frost' covering the surface that needs to be carefully removed; depending on the floral source of the honey, it can be up to 25 mm deep. This white frothy substance is dextrose hydrate. Fermentation may occur in this layer following granulation. When dextrose hydrate separates, the remaining honey increases its moisture content. The surface layer exposed to the air will absorb even more moisture and can become liable to fermentation.

METHOD A

Bulk stored honey that has granulated with a fine set crystal may be processed by warming sufficiently without liquefying, stirring to a smooth consistency and immediately bottling into warm jars.

METHOD B

Bulk stored honey with large coarse crystals requires special treatment that includes totally reliquefying and then seeding with honey of a fine crystalline texture.

- Reliquefy each 15 kg bucket in a warming cabinet that is thermostatically controlled. Check regularly during

- the heating process which can take up to eight hours or longer. It is essential that the honey is not overheated.
- When the honey is melted and there are no crystals present, allow it to cool to about 30 °C (86 °F).
- Carefully warm three 454 g (1 lb) jars of good quality fine set honey sufficient to soften but not to liquefy it. Carefully add it to the 15 kg of bulk honey, mix thoroughly but do not aerate it. The use of a honey creamer makes incorporation so much easier.
- When you feel the honey is adequately blended, stand the container in the warming cabinet at bottling temperature for several hours (trial and error), until any air bubbles have risen to the surface. Bottle.

SEED HONEY

To prepare a quantity of seed honey, some of the previously seeded honey can be stored in a bulk container, labelled accordingly and stored in a cool, dry place. This honey should be set in three to four weeks, producing fine-grained honey that can be used to process further soft-set honey for bottling.

A METHOD USING LING HEATHER HONEY

Ling heather added to a bland honey such as oilseed rape will give the finished product a tan-coloured appearance. Removing the jar lid enables the honey to give off an immediate aroma of heather honey that is most pleasing.

- The container of bulk (non-heather) honey should be heated in a warming cabinet to 40–45 °C (104–113 °F)

Appendix C: A Method of Producing Soft-set Honey

for six to eight hours, until the mass softens. It should not be allowed to liquefy.
- The bulk should 'mashed' with the use of a creaming device, avoiding the introduction of air at all costs. The treated honey should be free of lumps and resemble a thin gruel.
- Add 1.36 kg (3 lb) of good quality ling heather honey, ensuring that it is thoroughly incorporated into the bulk. The ling heather honey should be as pure a sample as is available and totally free of any granulation.
- Once the mass of honey is homogeneous throughout, it can be bottled immediately into jars that are best slightly warmed.
- When filled, each jar should be placed on a level surface while it sets; it is a most pleasing sight to see a level surface when the lid is removed from a jar to reveal a smooth, dry, level surface, characteristic of a quality soft-set honey.

APPENDIX D: VARROA CONTROL FOR HEATHER STOCKS

There many schools of thought on how to control varroa in stocks going to the heather moors. Contact the following organisations for advice:

England and Wales:

>National Bee Unit
>The Animal and Plant Health Agency (APHA)
>National Agri-food Innovation Campus
>Sand Hutton
>York
>YO41 1LZ
>
>Telephone: +44 (0)300 303 0094
>E-mail: nbu@apha.gsi.gov.uk
>Website: www.nationalbeeunit.com

During the season, support will also be available locally from the seasonal bee inspector for your area.

Scotland:

>Science and Advice for Scottish Agriculture (SASA)
>Roddinglaw Road
>Edinburgh
>EH12 9FJ

Appendix D: Varroa Control for Heather Stocks

Telephone: +44(0)131 244 8890
E-mail: info@sasa.gsi.gov.uk
Website: www.sasa.gov.uk

Northern Ireland:

Department of Agriculture, Environment and Rural Affairs (DAERA)
Dundonald House
Upper Newtownards Road
Ballymiscaw
Belfast
BT4 3SB

Telephone: +44(0)300 200 7852
E-mail: daera.helpline@daera-ni.gov.uk
Website: www.daera-ni.gov.uk

Republic of Ireland:

The Department of Agriculture, Food and the Marine (DAFM)
Agriculture House
Kildare Street
Dublin 2
DO2 WK12

Telephone: +353(0)1 607 2000
E-mail: info@agriculture.gov.ie
Website: www.agriculture.gov.ie

INDEXES

GENERAL INDEX

A Journey into Greece 59
Abbott Bros (Dublin) 168
access (site) 103, 246
acid
    acetic 95, 96
    amino 96
    butyric 96
    citric 95
    formic 95
    gluconic 96, 350
    glutamic 96
    malic 95, 96
    organic 77, 96
    succinic 95, 96
afforestation 30
agriculture, intensive 41
alighting board 241; **64**
alpine 19, 75
altitude 21, 73, 75, 99, 108–10, 246, 311; **115**
American Bee Journal 194
amphibian 40
Angle 50–51
Annulosa 65
antioxidant 96
aphid 112
Apidae 65
Apis 65
    cerana indica 66
    dorsata 66; *see also* giant honey bee
    florea 66; *see also* little honey bee
    mellifera 66, 143; *see also* western honey bee
        carnica 137
        ligustica 137
        macedonica 137
        mellifera 137, 139, 143–4, 295; *see also* north-European dark bee
        'Britannica' 128
        scutellata 66
aroma
    (honey) 98, 99, 306
    (wax) 287
Art of Beekeeping 18, 173
Arthropoda 65
aspect 115–6
auroch 48, 49
bacterium 350–1, 352–3, 356
basswood 275
beater 46
bedrock 114
bee 65
    aggressive 137
    Buckfast 140, 141, 143
    Caucasian 127, 325
    feral/wild 67, 68, 70, 170, 208, 292, 293–5; **62**, **293**
    Greek (Carniolan type) 143
    honey 65
        Asian (*Apis cerana*) 66
        giant (*Apis dorsata*) 66
        little (*Apis florea*) 66
        western (*Apis mellifera*) 66

    hygienic 123
    north-European dark 139
    solitary 65
bee bole 56–7; **57**
Bee Craft 60, 124, 342
bee farm 202; **61**; *see also* beekeeper, commercial
bee house 138; **138**
Bee Improvement and Bee Breeders' Association (BIBBA) 128; **129**; *see also* Village Bee-Breeders' Association, British Isles Bee Breeders' Association
bee inspector 234
bee space 57, 150–2, 159, 160, 161, 162, 170, 217, 218, 286; **151**, **214**
beekeeper, commercial 60, 137, 140, 156, 161, 166, 194, 238, 264, 266, 281, 320, 331, 338; **104**, **302**; *see also* bee farm
    semi- 322
Beekeepers' Record 60
Beekeeping in Britain 24, 336–7
Bee-Keeping New and Old, described with pen and camera 321
Beekeeping Techniques 18
Beekeeping Up-to-date 18
Bees: Their Natural History and General Management 320
Bees to the Heather 17
Beeswax 50, 51, 53, 56, 59, 60–1, 172, 173, 276, 279, 285, 286; **61**
bee-way 273, 276, 279
benweed (*Senecio jacobaea*) 110
bilberry (*Vaccinium myrtillus*) 40, 74
biodiversity 40
bird; *see also* capercaillie, grouse, partridge, pheasant
    game 39, 41
    ground-nesting 33
bird life 38
blackberry (*Rubus fruticosus*) 112, 206, 208, 220
Blackburn Aircraft Company 327, 337
blanket bog 29, 79, 91; *see also* bog
boar 48
bog 74; *see also* blanket bog
borage 140, 183, 206, 208, 220
boreal forest 42
bracken 30, 32, 33–5, 40, 49, 87
    spraying 33; **33**
breakdown (vehicle) 103
Breeding the Honeybee 143
British Bee Journal 60, 169, 321, 329, 339
British Bee-Breeders' Association 123–4
British Beekeepers' Association (BBKA) 17, 24, 59–60, 158, 335
British Isles Bee Breeders' Association 128; *see also* Bee Improvement and Bee Breeders' Association
British Standard 152, 154, 158, 159, 161, 166, 212, 341
British Standards Institution (BSI) 24, 341
brood
    cell 160, 175, 176, 201; **175** *see also* cell (honey)
    drone 133, 173

374

Indexes

worker 133, 174
chilled 215
emerging 222, 223, 229, 230
rearing 184, 198, 207, 208, 209, 220–1, 227, 243
sealed 181, 193, 200, 205, 206, 212, 213, 216, 217, 218, 230; *231*
surplus 211
unsealed 195, 199, 229, 230, 312
broom 76
bubble, air 101, 261, 262, 304; *94*, *100*, *308*
buffer 95
bumblebee 65
burning 29, 32, 33–7, 38, 46, 48, 49, 105; *34*, *35*; *see also* swailing
Burtt & Son 159, 323, 331, 332–6
Butler cage 223; *see also* queen cage
calcium 96
Calluna vulgaris 19, 40, 72, 74, 348; *72*; *see also* ling heather
candle 53, 54
candle making 54
capercaillie (*Tetrao urogallus*) 42
capping 253, 255, 262, 264, 297; *see also* uncapping
carbon 39
cell (honey) 99, 172; *see also* brood cell, queen cell, swarm cell
centrifugal force 99, 171; *171*; *see also* extraction
centrifuge 101, 244, 245, 260–3, 264; *261*; *see also* extractor, spin dryer
ceresin 56
Chain Bridge Honey Farm *264*, *299*
chalkbrood 215
Chamerion angustifolium 110; *111*; *see also* fireweed, rosebay willowherb
charter 56
cheesecloth 263
chlorine 96
class
classification 65
(exhibition/show) 289, 306
clearer board 246–7
Canadian type 246
Forrest 247; *247*
Cleveland Show 101
climate 29, 75, 138, 203, 270, 281, 336
change 41, 42, 47, 93, 102, 189
clover 108, 140, 183, 184, 206, 208, 220, 221, 225–7, 272, 319, 325, 329, 343
honey 368
cluster 186, 209, 271, 297
cog 232, 284; *see also* eke
collar (guide) 306, 310; *307*
colony
cohesion 135, 174
feral/wild: *see* bee, feral/wild
strength 21, 175, 201, 206, 220–9, 271, 290–3
population 21, 179
resources 184
colour (honey) 19, 95, 98, 99, 113, 308, 370; *270*, *304*, *308*
comb
arrangement 178, 185, 202, 203, 230
brace 178, 248, 251, 279
brood 154, 175, 177, 184, 200, 203, 216, 224, 229, 286; *175*, *230*

building 179–82; *180*; *see also* nucleus, comb-building
burr 178, 248
change, Bailey 297
drawn 154, 169–83, 232, 248, 251, 263, 273, 276, 281, 286, 292
honey 169, 171, 172
replacement 172–4
sealed 181, 229, 232, 294; *see also* brood, sealed
unsealed 232; *see also* brood, unsealed
commando training 36
composition (honey) 98
conifer 38
plantation 42
conservation group 39
constituent (honey) 94–5, 96
antibacterial 355
mineral 95
container
bulk 252, 262, 263, 266
cut-comb 284, 287–9; *288*
contamination (honey) 99
Court of Alderman 54
cowl 257; *258*
crofter 29
crop (honey) 140
heather 20, 21, 59, 102, 114, 238, 243, 246, 248, 271, 292, 323; *122*
late 108, 183, 208, 220–9; *221*, *223*
crownboard 161, 163, 181, 184, 185, 186, 193, 209, 210, 224, 232, 236, 239, 240, 260, 272, 291, 297
crystal 98; *see also* granule
crystallisation 98, 99, 172; *see also* granulation
custodian 32, 39
damping down *34*
day length 21, 206, 312
daylight 164, 237, 238
deer 48, 49
stalking 46
deforestation 41; *36*
Demaree; *see also* swarm control, swarm prevention
method/system 23, 183, 189, 194–201; *196*, *197*
reverse 199–200; *197*
density (honey) 96–7, 98
Department for Environment, Food and Rural Affairs (Defra) 39
deoxyribonucleic acid (DNA) testing 48
desensitisation (pollen) 356
dextrose 96; *see also* glucose
diet (grouse) 39, 41, 43
diploid 130
disease 124, 125, 257–9
control 179
Isle of Wight 336
transmission 241
Dissolution of the Monasteries 44, 56
ditch 39
divider (section) 273; *278*
dog 44, 46, 48, 49
handler 46
dogging 46
Domesday Book 51
doubling 183–7, 220, 222
Douglas fir (*Pseudotsuga menziesii*) 112

375

dressing 351, 352, 354–5, 356
drifting 193, 241
drone 130, 131, 132–6, 141, 145, 146, 174, 177, 187, 188, 195, 199, 209, 212, 215, 216, 296; *see also* brood cell, drone; queen, drone-laying; worker, drone-laying
drought 75, 77, 93, 102, 105, 246
drumming 53
dry stone wall 240; **240**
dysentery 172, 292
ecology 44, 50, 76–9
ecosystem 31
egg 130–1, 175, 176, 181, 195, 200, 202, 205, 229, 230, 271; **231**
eke 181, 232, 247, 284, 286; **165**, **285**; *see also* cog
element
    (heating) 253, 267
    (mineral) 96
entrance 110, 161, 164–5, 166, 167, 168, 185, 189, 193, 194, 211, 213, 216, 217, 218, 223, 232, 233, 236, 239, 246, 252, 260, 297, 312; **165**, **214**
entrance block 164, 296; **296**
environmentalist 32, 77
enzyme 95, 269, 350
epoch 27
Erica 19, 73, 83–92; *see also* heather, bell
    carnea 72; *see also* heath, Mediterranean
    ciliaris 72, 83; *see also* heath, Dorset or Arn
    cinerea 40, 72, 74; **72**
    tetralix 72, 73, 74; **72**; *see also* heath, cross-leaved
    vagans 72; *see also* heath, Cornish
Ericaceae 75, 76
erosion 49
estate 28, 40, 44
evening primrose 183
evolution 63–4, 66
exhibition 59, 233, 303–10; *see also* show
extraction 99, 157, 171, 232, 251, 253, 254–5, 257, 269
    centrifugal 98, 245, 262
extractor (honey) 331; *see also* centrifuge
    parallel-radial 337; **328**
    radial 156, 171, 261, 264; **171**
    tangential 99, 171, 255, 257; **171**, **256**
family (classification) 65; *see also* Ericaceae
farmer 31, 35, 53, 103, 157, 167, 338
farmers' market 298
farming 50, 325, 329
    sheep 28, 29, 325
farmland, scrub 110–11; **111**
fastener (hive)
    clamp **236**
    lock-slide 236
    spring 236; **235**
    staple 236
    strap 235, 238, 239, 240; **104**, **165**
    toggle 236
feed hole (crownboard) 209, 210, 224, 297
feeder 163, 185, 209, 213, 218, 222, 223
feeding (bees) 135, 144, 185, 186, 199, 209, 213, 222, 228; **344**; *see also* syrup (sugar)
fermentation 101, 263, 266, 269, 292
fireweed (*Chamerion angustifolium*) 112
flavour (honey) 73, 95, 98, 102, 270

floor/floor board 161, 163, 166, 176, 177, 185, 193, 209, 210, 225, 252, 292, 318; **177**
    debris 296
    heather 168; **169**, **237**
    open-mesh 294
    solid 294
flora 32, 108, 290
flow 97, 101, 303, 306, 307
    honey/nectar 21, 105, 156, 178, 179, 183, 192, 195, 198, 204, 205, 220, 221, 222, 225, 226, 242, 243, 251, 270, 271, 284
foam strip 164, 236, 241; **165**
fog 108
foliage 102, 105, 106
fondant 209
food source (grouse) 40
forage 109, 138, 204, 220, 281, 325, 343
forager/foraging bee 67, 106, 108, 149, 173, 184, 188, 189, 193, 198, 204, 205, 206, 216, 220, 221, 233, 291
    pollen 213; **64**
foraging force 102, 188, 205–7, 209, 211, 229, 295
forest 32, 40, 42, 47, 51
forestry 38, 93, 110–1; **111**
Forestry Commission 93, 110
Forestry Enterprise (England) 40
foundation (beeswax) 156, 172, 173, 175-6, 177, 178–9, 181, 185, 189, 193, 195, 198, 199, 201, 202, 203, 210, 216, 217, 222, 224, 228, 232, 245, 248, 276, 280, 282, 283; **249**, **250**
    starter 135, 232, 245, 275, 276, 285; **287**
    thick 275
    thin 251, 275, 285
frame 153
    lug 154, 336
        long- 158, 159
        short- 154, 158, 160, 161, 162
    Manley 156; **156**
    movable 59
    runner **158**
    spacer 154–7; **157**
    spacing 153, 154–7, 178, 248–9; **249**
    Hoffman 155; **155**
freezer 99, 289, 307
frost 57, 109, 115, 243, 246, 290
    damage 106
frothing 261, 269
fructose 96; *see also* levulose
furze 76
gamekeeper 31, 103
gene pool 136
General Apiarian 317
genetics 125, 126, 130, 132
genotype 131, 132
genus 19, 65, 75
geography 70, 75
giant redwood 93
Gleanings in Bee Culture 280
glucose 96, 350; *see also* dextrose
glucose oxidase 350
goat 48, 49
government agency 31
grading up 130
granulation 112, 289, 307; *see also* crystallisation
granule 112, 306, 308; *see also* crystal
grassland 28, 29, 48, 49, 51

376

acid 40, 74
  marginal 29
Great Yorkshire Show *129*
greenstone 115
ground condition 103, 112
grouse 28, 30, 32, 35, 39, 40–7, 74, 105, 114, 311
  ancestor 42
  black (*Tetrao tetrix*) 32, 39, 41
  red (*Lagopus lagopus scoticus*) 39, 41, 45; *43*
  sustenance 41
  willow 41
habitat 30, 32, 40, 42, 74, 76, 116
Handy Book of Bees 320
haploid 131
hardwood 36
  forest 47
Hartpury College *52*
hawthorn 206
heath 20, 49, 72, 73, 75, 76, 77, 83–92
  Cornish (*Erica vagans*) 72, 83
  cross-leaved (*Erica tetralix*) 72, 73, 74, 82–92, 98; *80*
  Dorset or Arn (*Erica ciliaris*) 72, 83
  Mediterranean (*Erica carnea*) 72
heath beekeeper 321-2
heather
  bees 139–40, 293
  beetle (*Lochmaea suturalis*) 102, 105–6
  bell (*Erica carnea*) 19–20, 30, 40, 72, 73, 74, 98; *20*, *26*, *80*, *81*, *316*
  ling (*Calluna vulgaris*) 19, 35, 39, 40, 41, 72, 73, 74, 75, 78, 98, 108, 114–5, 221, 343
  plant 29, 32, 35, 49, 72-3, 105, 115, 116, 311, 312
  press 140, 264
    Mountain Grey (MG) 266, 331; *265*
    Peebles 266; *264*
  seeds 35
  young 32, 35, 41, 105
heather-going 18, 21, 24, 108, 317
heathland 20, 74, 75, 76-9, 83–5, 321
Heddon method of reinforcement 189, 192; *191*
helicopter 33
hen harrier 39
Highland clearances 29
Himalayan balsam 290
History and Management of Bees 320
hive
  movable-frame 56–9; *58*
  Baker 166; *166*
  Burtt's of Gloucester 159
  catenary 164
  Commercial (Modified) 151, 159, 161
  Congested Districts Board (CDB) 167–8, 335; *167*, *168*
  cottager 53, 57
  Dadant 281
  Dartington 163
  Deep National (14x12) 160, 163, 164
  double-walled 150, 336; *see also* WBC hive
  Glen Imperial 167; *167*
  Langstroth 150, 156, 159, 162, 163, 164, 251, 281
  leaf 59
  log 53, 70
  Mountain Grey (MG) 24, 323, 336–41; *340*
  National (Modified) 151, 155, 158, 159–60, 323, 334, 335–6
    British Standard 24, 152, 158, 159–60, 341; *340*
    Improved 323, 336–341; *see also* Mountain Grey (MG) hive
  polystyrene 150, 153, 159, 163; *104*
  Simplicity 159; *see also* Commercial hive
  single-walled 150, 153, 161, 335
  Smith 150, 158, 159, 160–1, 176, 183, 184, 292
  top-bar 163
  Warré 163
  WBC 53, 57, 150, 166, 167, 176, 183, 184, 336
Holocene epoch 27, 47, 67
Homo sapiens sapiens 67; *see also* man, early; man, Mesolithic; man, stone age
honey
  bell heather 20, 98, 183–203; *99*
  blend(ed) 21, 98, 108, 112, 204, 269–70, 304, 306; *270*
  comb 99, 246, 275, 281
  cut-comb 99, 233, 240, 245, 279, 280, 284–9; *275*, *285*, *286*, *288*, *299*
  extracted 248–9
  floral 95–8, 99, 156, 160, 207, 251, 350, 352, 355
  granulated 251–3
  ling heather 20, 94, 95, 99–101, 105, 242, 260, 266, 269, 352; *94*, *100*, *301*, *305*, *308*
    exhibition 303–10; *310*
    pressed 263–6
  oilseed rape (OSR) 172, 251, 269
  section 139, 148, 232, 233, 240, 245, 270–9, 280, 281, 289; *274*, *277*, *278*
honey hunter/hunting 51, 67, 68–9, 70; *68*, *69*
honeydew 112–3
honeysuckle (*Lonicera* spp.) 62
horse chestnut 47, 206
Horsley board 189, 208, 224; *see also* swarm control
human activity 41
hunter-gatherer 66, 70
hydrogen peroxide 255, 350
hydroxymethylfurfural (HMF) 266, 269
hygroscopic 266, 292, 362
hymenoptera 65
ice age 19, 28, 47
import (bees) 66, 123, 124–5, 127, 128, 138, 139, 336
incentive, financial 32
industrial activity 41
Industrial Revolution 44
infection 351, 353, 355, 356
insect 40, 41, 63, 64–5, 145, 271, 284
instrumental insemination 130, 131, 134, 141
insulation 153, 163, 164, 185, 186, 243, 272, 273
interglacial period 27, 28
International Bee Research Association (IBRA) 57
intervention 28, 76
ivy 290
Ixodes ricinus 25; *see also* sheep tick
Japanese red cedar 93
jar (honey) 70–1, 97, 98, 101, 262, 269, 270, 300, 304–6, 309–10; *94*, *299*, *305*, *307*
judge, honey 23, 98, 99, 191, 113, 289, 303, 304,

306, 307, 308, 309, 310; ***308**, **310***
judge's mark/strike 101; ***305***
Jutes 50–1
Kent Beekeepers' Association 60
Killion board 176; ***177***
kingdom (classification) 65
label 279, 283, 284; ***299**, **301***
    class (exhibition/show) 306, 309, 310; ***307**, **310***
Lagopus lagopus scoticus 41; *see also* grouse, red
Lagopus mutus 42; *see also* ptarmigan
Laidlaw Bee Research Centre 131
landowner 29, 32, 38, 43, 45, 103, 110, 240
larva 145, 146, 175, 176, 181, 205, 212, 230; ***231***
leaching (soil) 77; ***77***
leaf mould 77–8
levulose 96; *see also* fructose
lid 101, 306, 309
    snap-on (bulk container) 266
    snap-on (cut-comb container) 284
lifespan
    (bee) 207
    (plant) 35, 76
lightning 48, 49, 67
lime 47, 206, 319
    wood 275, 281
livestock 48, 49, 50, 54, 130, 343; ***104***
Lochmaea suturalis 105; *see also* heather beetle
Lonicera spp. 62; *see also* honeysuckle
lysine 96
making increase 215–9
maltose 96
man
    early (*Homo sapiens sapiens*) 62, 67, 68, 70; *see also* man, Mesolithic; man, stone age
    Mesolithic 68, 70; *see also* man, early; man, stone age
    stone age 48; *see also* man, early; man, Mesolithic
management, chequerboard 202
Margaritifera margaritifera 40; *see also* mussel, freshwater pearl
mating 126, 130, 131, 133, 134, 137, 144; *see also* instrumental insemination; nucleus, mating
    apiary 131, 133, 134, 141, 145; ***142**, **146***
    hand 131
mattgrass 40, 74
Meleagris gallopavo 42; *see also* turkey, wild
Mellifica lehzeni 143
metal end 154, 157; ***249***
methylated spirits 310
Meyers Timber 327, 329, 339
midrib (comb) 154, 174, 260
mineral 56, 77, 78, 95, 96
Miocene epoch 42
mire 74
mist sprayer 237
mobile phone signal 103
moisture 115, 292, 350, 354
monastery 53, 59, 140; *see also* Buckfast Abbey
monk 49, 51, 140; *see also* Adam (Brother)
moor/moorland; *see also* muir
    custodian 32, 39
    grouse 32, 39, 43, 46, 105, 311
        licensing 47
    habitat 30, 32, 40, 74, 116
    heather 16, 21, 27, 28, 30, 31, 32, 48–9, 74,

81, 83–92, 102, 112–3, 150, 229, 233, 239, 284, 300, 311, 312, 317, 347; ***20**, **22**, **26**, **31**, **74**, **80***
    management 29, 31–2, 46, 300
    upland 18, 20, 28, 30, 31, 32, 38, 40, 41, 43, 45, 49, 51, 73, 74, 114, 207, 352
moorkeeper 103, 114
Moorland Association 30, 39
Mountain Grey Apiaries (MGA) 123, 133, 159, 184, 323–332, 341
mouseguard 208, 296
muir 76; *see also* moor
muslin 303
mussel, freshwater pearl (*Margaritifera margaritifera*) 40
mutation 75
National Honey Show 303, 345; ***343***
national park 41; ***26**, **188***
nectar 21, 65, 94, 95, 98, 105, 108, 110, 112, 115, 116, 178, 179, 184, 185, 186, 187, 199, 205, 220, 222, 225, 227, 228, 230, 248, 271, 281, 282, 290, 311, 348, 350; *see also* flow, honey/nectar
    dearth/shortage 149, 257
    source 206, 343
neglect (land) 30
Normans 27, 43, 49, 51
northern hemisphere 42
nosema 297
nucleus 173, 174, 201, 205, 206, 208, 210–11, 212, 222, 223–4, 228, 229, 232, 233, 290–1, 292, 294
    comb-building 179–82; ***180***; *see also* comb building
    mating 132, 141
    method 212–9
    hive ***214***
    mini 145–6; ***146**, **147***
nurse bee 135, 176, 182, 187, 202, 212, 213, 219, 224, 227, 247
nutrient 49, 95
nutritionist 96
oak 32, 47, 93
odour 148, 223, 273, 310
    hive 217, 224
oilseed rape (OSR) 179, 184, 206, 220, 220, 251; *see also* honey, oilseed rape
organic substance 95
orientation (hive) 54, 192, 193
overgrazing 30
overheating
    (bees) 237
    (honey) 303, 306
overwintering 184, 294, 343; ***296***
Pagden method 189–94; *see also* swarm, artificial
pan
    humus 77
    iron 77
parthenogenesis 130, 134
partridge 41, 42
pasteurisation 101, 266
pasture 110, 272, 343
peat 76, 77, 78, 79, 105; ***79***
Perforextractor 254; ***255***
phacelia 183
pheasant 41, 50

Indexes

pheromone 68, 148, 195, 201, 224
phylum 65
Picea sitchensis 40; *see also* Sitka spruce
Picea spp. 112; *see also* spruce
pine 36, 42, 93
Plantagenet dynasty 51
plantation 42, 93, 110–2; *113*
podzol 77–8; **77**
pollen
    analysis 48
    arch 177, 295
    ling **64**
    mite 172
pollination 338
polystyrene 150, 153, 159, 163; **104**
Porter bee escape 247
potash 48, 112
potassium 96
Pratley uncapping tray 252
predator 31, 68, 163, 296
press, beekeeping 127
press, heather: *see* heather press
pressing 245, 260, 263; **265**; *see also* heather press
proline 96
propolis 149, 150, 157, 158, 172
propolisation 152, 154, 158, 168, 181; **158**
protein 96, 307
    colloid 94
Pseudotsuga menziesii 112; *see also* Douglas fir
ptarmigan (*Lagopus mutus*) 42
purity (honey) 99, 100, 204, 307, 308; **305**
queen
    cage 219, 223; *see also* Butler cage
    cell 54, 139, 144, 146, 149, 192, 193, 198, 199, 200, 201, 202, 208, 209, 212, 213, 216, 218, 219, 223, 228; *see also* cell (honey)
        emergency 173, 181
        grafted **146**
        supersedure 202
        surplus 192, 212
    clipped 188
    drone-laying 297
queen excluder 163, 167, 173, 179, 186, 187, 193, 195, 199, 222, 225, 227, 281, 282, 291, 295; **286**
queen substance 202
quilt 209
ragwort (*Senecio jacobaea*) 110; ***111***
rain 57, 108
    fall 75, 311
    water 77; **107**
'Red List' species 42
red-flowering currant (*Ribes sanguineum*) 189
refrigerator 99, 309
    discarded 266
regeneration 32, 35, 49
research, archaeological 49
Rheology of Honey 94
rhizome 33
Ribes sanguineum 189; *see also* red-flowering currant
right of way 108
rod, glass 101, 307
rodent 241
Romans 27, 50, 351
rosebay willowherb (*Chamerion angustifolium*) 110, 204, 206; ***111***

Royal Society for the Protection of Birds (RSPB) 46
Rubus fruticosus 112; *see also* blackberry
runner, frame *158*
run-off 40
St James's wort (*Senecio jacobaea*) 110
Saxon(s) 50–1
    law 44
Scottish Beekeeper 126
Scottish Beekeepers' Association 28
scrape 39
scratching post 105, 312
scrim 262, 263, 264, 303; **263**, **265**
sea fret 108
seal
    (document) 50
    (lid) 309
seasonal variation 245–6
section
    bait 273
    basswood 232, 273
    circular 279–83
        Cobana 245, 279
        Ross Round 99, 232, 245, 279–283; **282**, **283**
    crate 233, 246, 270, 273, 274, 281, 282; **274**, **282**
    frame 272, 276, 282
    jig 276; **277**, **278**
    rack *278*
sedge 74
sediment 40
Senecio jacobaea 110; ***111***; *see also* benweed, ragwort, St James's wort, staggerweed, tansy ragwort
sheep 28, 29–30, 50, 51, 103, 105, 312, 325; **80**
    tick (*Ixodes ricinus*) 35
sheep's fescue 40, 74
shelter 106, 110, 170; **26**, **107**
    bee **58**
shepherd 31, 48, 103
shoots, young (plant) 29, 42, 312
shotgun, breech-loading 46
show 23, 101, 279, 303, 306, 308, 309, 345; **129**, **310**, **343**; *see also* exhibition
    schedule 309; **307**
shrub 19, 30, 74, 76
site 102–16, 243, 311; **106**, **107**
Site of Special Scientific Interest (SSSI) 30, 74
Sitka spruce (*Picea sitchensis*) 40
Sjolis honey loosener 255; **256**, **258**
skep 51, 53, 54, 56–7, 59, 189; **52**, **58**
    apiculture 51
    beekeeper/skeppist 54, 280, 320
smoke, use of 67–8, 71, 179
Snainton Bee Farm **61**
Snelgrove board 189, 224; *see also* swarm control
snow gully *109*
soil 49, 73, 75, 77–8, 98, 99, 112, 114–5, 311
Special Areas of Conservation 30
Special Protection Areas 30
species (classification) 66
specific gravity (SG) 96–7
spin dryer 260, 261–3; **263**; *see also* centrifuge
spraying, helicopter **33**
spruce (*Picea* spp.) 40, 93, 112, 113; *see also* Sitka spruce
staggerweed (*Senecio jacobaea*) 110

379

stakeholder 38–40, 47, 103
stance 106–16, 204, 205, 235, 238, 239, 240–1; *104*
stearin 56
Steele & Brodie 168, 333; *167*, *168*
sting (bee) 148–9, 257
stock improvement 130–7
straining cloth 303, 304
strap (hive) 232, 235, 238, 239, 240; *104*, *165*
stretcher, carrying *244*
strike mark *305*, *308*
sub-kingdom (classification) 65
sucrose 96
sugar 59, 95, 96
    concentration 356
    rationing 334, 338, 341, 342–5; *344*
sulphur 96
super 156, 157, 165, 178–9
swailing 32; *see also* burning
swarm
    artificial 189, 293; *190*, *191*; *see also* Pagden method
    cast 192, 193, 210, 216, 218, 219, 294
    cell 179, 189, 195, 224; *see also* cell (honey), queen cell
    control 54, 187–9, 194, 224; *see also* Demaree, Horsley board, Pagden method, Snelgrove board
    'forced' 208–10
    impulse 199, 201, 202
    prevention 194, 198, 200, 205; *see also* Demaree
sycamore 47, 206, 319; *62*
syrup (sugar) 185, 186, 209, 213, 220, 222, 223, 292; *see also* feeding
tallow 54
tansy ragwort (*Senecio jacobaea*) 110
taste 95, 98, 99
tax 38, 53
    break 38
temper 132, 137, 148
temperament 139, 148
    group 137
temperature
    (hive) 155, 164, 176, 181, 236, 237, 248, 273
    (honey) 266, 267, 269, 289
tenurial law 29
terrain 103, 246
Tetrao tetrix 41; *see also* grouse, black
Tetrao urogallus 42; *see also* capercaillie
Teucrium scorodonia 112; *see also* wood sage
thixotropy 94, 95, 99, 101, 255, 257, 305; *94*, *100*
timber 153, 276, 281, 284, 327, 329, 330
    licence 329, 331
trailer 236, 238; *239*
trait 71 130, 131, 132, 135, 140, 143, 148, 336
turkey, wild (*Meleagris gallopavo*) 42
uncapping 156, 172; *see also* capping
uniting 165, 210, 211, 214, 229–30, 232, 233, 290
    newspaper method 291; *291*
University of Twente 93
Vaccinium myrtillus 74; *see also* bilberry
vandalism 108, 113

varroa 133, 138, 294
    treatment 234
vegetation 32, 49, 51, 75, 76, 78, 79, 114, 241; *78*, *107*
vehicle
    four-wheel-drive 103
    off-road 246
ventilation 165, 188, 233, 237, 312, 336
    screen 164, 236, 239, 247; *165*
vermin 289
Vietnam War 60–1
Vikings 27, 50–1
Village Bee-Breeders' Association 128; *see also* Bee Improvement and Bee Breeders' Association
viscosity (honey) 95, 96–7, 98
vitamin 41, 96
warming cabinet/device 252–3, 266–8, 303; *267*, *268*
wasp 259
water
    content 94, 101, 263, 269
    shortage 102
water vole 40
watercourse 40, 106, 290
waterproofing 50
wavy hair grass 74
Wax Chandlers' hall *55*
wax extractor
    MG 331; *330*
    solar 173, 263, 264
wax melter 252; *254*
    MGA 263
wax moth 172
Weeds Act 110
weight
    (hive) 160, 162, 167, 248, 295, 318, 319
    (product) 283, 308
western red cedar 153, 327, 341
whinstone 115
wildlife 30
wind 30, 106, 110, 115, 116, 144, 148, 246; *26*, *104*, *107*
wing venation 123, 295
winter preparation 290–3
wood sage (*Teucrium scorodonia*) 112
woodland 32, 36, 38, 47, 48, 78, 93, 112
woodpecker 163, 296
Woolton Pie 342
worker, drone-laying 297
World War I/II 36, 46, 254, 272, 338, 341–9, 351, 352; *36*
Worshipful Company of Wax Chandlers of London 54–6
writing tablet 50
yeast 95, 101, 262, 266, 269
York Bee Farms 202
York method 202–3
Yorkshire Apiaries 123, 184, 331; *236*, *255*
Yorkshire Beekeepers' Association (YBKA) 339
Yorkshire County Beekeepers' Association 339
Younger Dryas 47
zinc 164

Indexes

PEOPLE

Abbott, Arthur 60, 123–8, 133, 159, 323–31, 337–41; *332*
Adam (Brother) 108, 134, 139, 140–44; *142*; *see also* Kehrle, Karl; monk
Albert (Prince) 28, 46
Allen, Harry 169
Allen, M Yate (Reverend) 325–7
Allen, Walter 331
Aristotle 355
Armitt, Harold 21
Aston, David 17
Austin, Charles 202
Badger, Jack 'Tim' 23, 351
Badger, Michael *310*
Barker, Jack *129*
Bielby, Bill 144, 295
Birch, Daphne *343*
Bloomer Cooper, T 280
Bradford, Tom 22, 260
Brandstrup, Keld 137
Brown, Tom 139
Burtt, Edward 333
Burtt, Michael 323
Butler, Colin 162, 195
Carr, William Broughton 59, 60
Churchill, Winston 342, 343
Clemmitt, William 22
Cooper, Beowulf 128, 133, 137, 139; *129*
Cordingly, David 17
Cowan, Thomas 60
Craig, Ian 294
Crane, Eva 18
Cromwell, Thomas 56
Davies, Glyn 145
de Bracey Marrs, Graeme 23
Deans, Alexander 18, 114
Denton, Frank 22
Dews, John *129*
Dioscorides, Pedanius 351, 355
Duke of Sutherland (first) 28
Eade, Brian 322
Eade, John (Eddy) 133, 322, 331, 332
Earl of Peel (third) 44
Ellis, JM 139
Flatman, Ivor 285
Foxton, Donald *129*
Gale, AW 60, 126
Grainger, Harry 27, 165
Green, George 339
Hamilton, William 18, 102, 114, 173
Hardy, Peter 22
Harold (King) 43
Hawes, Alan 22
Hebden, Alfred 121, 322
Henderson, Sammy *166*
Herrod–Hempsall, Joseph 60
Herrod–Hempsall, William 60, 320, 335–6
Hind, Albert 321
Home, John *302*
Hopwood, Charley 22
Huber, François 59
Hughes, J Eric 22, 144
Huish, Robert 320

Hyde, Austin 342–5; *343*
Isaacs, Jacob 317
Jackson, Ken 133, 331
Jenkin, Phil 22; *142*
Jesper, Dennis 22
Jones, Gavin 138
Kehrle, Karl 140; *see also* Adam (Brother)
Langstroth (Reverend) 57
Lawson, Nigel 38
Leafe, Bernard 22, 308
Ledgard, Sammy 322
Leng, George 22
Lister-Kaye, John 28
Lockwood, Bill 23
Lumley, Phil *129*
Mabane, William 345
Madoc, EWD 60
Manley, ROB 24, 60, 336
Marshall, Ged 137
McScott, Peter 262
Miller, Robert Warren 44
Mills, John 139
Moxon, Gerald 144
Osborn, Erica 342
Pearson, Terry 22, 104
Pettigrew, A 320
Phillipson, Martin 22
Pierson, Herbert 112
Prokopovych, Petro 280
Pryce-Jones J 94
Rawlins, Anthony 338
Reed, Robert 280, 320
Renshaw, Jack 144
Reynolds, Bill 23, 241
Roantree, Arthur *332*
Robinson, Dennis 22; *343*
Salmon, Walter 22
Scaife, Arthur 166
Schollick, Peter 17
Seeley, Tom 294
Settle, Maurice 23
Sims, Donald 19
Slinger, Will 23, 139, 229
Smith, Willie 161; *264*
Snowden, Paul 239
Sonley, Teddy 293
Spruce, Reg 144
Taylor, Ben *129*
Theaker, Terry 128, 133, 137
Tinsley, Joseph 18, 114
Tomlinson, Robin 22, 23, 194
Tonsley, Cecil 329
Victoria (Queen) 28, 46
Wadey, Jim 60
Weightman, Colin 22, 108, 139, 143, 280, 320, 337; *122*, *142*
Weir, John 93
Wheler, George 59
Whitehead, Stanley 17
Wighton, John 320
Wilkinson, Freddy 124, 133, 331, 338
Wilkinson, Hector 351, 352
William the Conqueror 43

381

Williamson, Arthur 23
Wood, Rowland 347
Woolhouse, Harold 130

Woolton (Lord) 342
York Jr, Harvey 135, 202

PLACES

Aberdeenshire 73, 89, 109; **20**, **115**
Aberfoyle 88
Aire Gap 85
Altnabreac 90
Anatolia 70
Apperley Bridge 345
Appleby-in-Westmorland 85
Arctic Circle 42, 73
Ardentinny 89
Argyll and Bute 89
Asia 42, 66
Attica 59, 355
Ayrshire 88
Babylonia 70
Bagshot Heath 83
Ballater **115**
Balmoral Estate 28
Banffshire 89; **81**
Barnard Castle 86
Bath 50
Battleton Farm (Kineton) 52
Beauly 28
Bicorp 68
Biggar 87
Birmingham 21, 39
Black Mountains 84
Blanchland 45, 73
Bleaklow 84
Bodmin Moor 83
Bog of Allen 90
borders (England) 87
Bradda 92
Bradfield Moor 73, 85; 80, 242
Braemar 109
Braemore 90
Brecknockshire 84
Brecon Beacons 84
British Isles 19, 20, 27, 28, 31, 41, 42, 47, 51, 60, 73, 74–93, 99, 102, 123, 128, 162, 183, 198, 203, 270, 279, 280, 281
Brora 89
Brough 86, 90, 322, 323, 324, 325, 337
Buckfast Abbey 60, 139, 140; **142**; *see also* monastery
Bulgaria 352
Buxton 84
Caernarvonshire 84
Caithness 79, 90
Cambus O'May 89; **115**
Canada 42
Cannock Chase 20, 73, 84; **113**
Cardiganshire 84
Carmarthenshire 84
Castle Douglas 88
Catterick 85
Cheshire 21, 84, 85
Cheviot Hills 86
Cleveland 81
Clyde 88

Comrie 88
Connemarra 90
Cooley Mountains 90
Cornwall 39, 76, 83
County Down 90
County Dublin 90
County Durham 73, 86; **58**
County Galway 90
County Kerry 90
County Louth 90
County Tyrone 90
County Wicklow 90
Crawford 87
Cregneash 92
Crianlarich 88
Crief 88
Cronk-y-Voddy 92
Crowle and Hatfield Moors 85
Crystal Palace (Sydenham Hill) 59
Culter Fell 87
Cumberland 86
Cumbria 85, 86, 139, 325
Cumnock 88
Dalby Moor 188
Dalby Mountain 92
Dallowgill 22, 85
Dark Peak 81, 84
Dartmoor 83, 134, 141, 319; **142**
Darvel 88
Darwen 86
Daviot 89
De Grey Rooms (York) 23
Denbighshire 84
Denton Moor 22, 322
Derbyshire 21, 81, 84, 85
Derry 90
Devon 39
Dinnet Valley 73, 89
Dorset 20, 73, 98; **99**
Dorset Heights 73, 83
Drummossie 89
Dubai 356
Dublin Hills 90
Dumfries and Galloway 88
Dunbartonshire 88
Dunnet Head 90
East Lothian 87
East Yorkshire (East Riding of Yorkshire) 20, 73, 81, 85, 86, 133, 144, 184, 323, 325, 332, 337; *see also* Yorkshire
Eden Valley 86
Edinburgh 86, 87
Egton Moor 35
Emley 85
England 30, 33, 39, 43, 46, 50, 51, 56, 87, 108, 127, 137, 140, 333; 82
Esk Valley 73
Ettrick Forest 87
Europe 42, 44, 47

382

Exmoor 83
Fairlie 88
Fewston reservoir 22
Findhorn Valley 89
Flow Country 79
Forest of Bowland 81, 86
Fountains Abbey 144, 295
France 49
Fylingdales 72, 80, 111
Galloway peninsula 88
Garelochhead 88
Garve 90
Glen Dhoo 92
Glen Moriston 89
Glen Truim 89
Gosforth 121
Great Britain 57, 72, 93, 321, 339
Greater Clydesdale 87
Greater Manchester 85
Greeba Mountain 92
Greenland 73
Greenwich 17
Gretna Green 88, 139
Gunnerside Estate 44
Hampshire 83
Hanover 73
Harrogate 85, 241, 262; *310*
Hartoft Dale 113
Haslingden 86
Haworth 85
Herefordshire 81
Holme-on-Spalding-Moor 113, 331
Hull 123, 234, 424, 331, 334, 338; *236*
Ilkley Moor 76, 85
India 67, 355
Ingleborough 86
Injebreck Colden 92
Inverness 28
Inverness–shire 36, 89
Ireland (British Isles) 41, 47, 57, 90, 167; *82*
Islay 89
Isle of Harris 138
Isle of Man 41, 81, 327; *91*
Jedburgh 87
Jordan Valley 355
Jorvik 51; *see also* York
Jura 89
Kennethmont 89
Kiev 280
Kildermorie 90
Kilmartin 89
Kincardinshire 89
Kineton *52*
King's Lynn 83
Kingston-upon-Hull 339
Kintyre peninsula 89
Kirk Ella 331
Kirkby Stephen *58*
Kirkcowan 88
Kirkcudbrightshire 88
La Arana 68; *68*
Lairg 89
Lake District 81
Lakeland fells 86
Lambfell Moar 92
Lammermuir Hills 76, 87

Lanarkshire 87
Lancashire 85, 86
Land of Nod *332*
Langholm 88
Latheron 90
Laurieston Moors 88
Leadhills 87
Leeds 130, 165, 194, 322, 339, 347; *138*
Leek 84; *22*
Lincoln 50
Lincolnshire 335
Lincolnshire Wolds 184, 325
Lingwell 90
Lizard peninsula 76, 83
Loch Earn 88
Loch Eck 89
Loch Kinord *20*
Loch Lomond 88
Loch More 89
Lochiel Old Forest 36
London 50, 54, 320, 342, 345; *55*
Lossie Valley 89
Lothian 87
Lower Foxdale 92
Luib 88
Lüneburger Heide 73
Lythe 129
Marlborough 126
Meikle Kinord 20
Melvich 89
Mennock Pass 88
Merionethshire 84
Mesopotamia 70
Middleton-in-Teesdale 85
Moffat 88
Moffat Hills 87
Moorfoot Hills 87
Moray and Nairn 89
Mount Hymettus 59
Mountains of Kerry 90
Mountains of Mourne 20, 73, 90
Muir of Dinnet 20
Muirsheil Regional Park 88
Mulgrave Castle *129*
Mull 89
Nepal 70; *69*
New Forest 83
New Galloway 88
New World 42
New Zealand 356
Newton Stewart 88
Norfolk 83, 388
North America 42
North Barrule 92
North Cave 133, 331
North Sussex Weald 83
North York Moors 93; *26*, *188*
North Yorkshire 44, 51, 73, 85, 86, 101, 139, 166, 184, 241, 293, 295, 322, 343, 347; *see also* Yorkshire
Northern Ireland 20, 73, 81, 91
Northumberland 45, 73, 81, 86, 87, 280; *122*, *264*
Oban 89
Old Malton *57*
Omagh 90
Ord of Caithness 90

Heather Honey: A Comprehensive Guide

Orkney 138
Oswaldtwistle 86
Outer Hebrides 138
Pateley Bridge 322, 347
Peebles 87, 266; *264*
Pennine chain 41, 81, 85, 86
Pentland Hills 87
Penzance 83
Perthshire 88
Picos de Europa 73
Princetown 142
Raasay 89
Radnorshire 84
Reeth 86
Renfrewshire 88, 294
Republic of Ireland 41, 81, 123
Richmond 85
Ripon 85; *343*
River Derwent 84
Rivington 86
Rombalds 85
Rosedale 113; *240*
Rossendale 86
Ross-shire 89
Rothbury 73
Rothiemurchus 36
Roundhay *138*
Roxburghshire 86
Royal Deeside *115*
Russia 42
Saddleworth 86
Saltersgate Moor 322; *114*
Sardinia 355
Sartfell 92
Scandinavia 42
Scarborough 61
Scargill 86; *58*
Scotland 28, 33, 41, 79, 81, 87, 102, 108, 161, 167, 235, 284, 320, 333; *82*
Scotcalder 90
Scottish Borders 86
Scottish Highlands 36, 42, 46, 76, 81
Scottish Lowlands 81
Seascale 121
Selkirkshire 87
Settle 85
Sherberton 134, 141; *142*
Shilford *122*
Siberia 73
Sixmilecross 90
Skipton 85
Skye 89
Slieau Curn 72
Slieau Freoaghane 92
Slieau Managh 92
South Africa 67
South Barrule 92
South Cave 325, 339

South Yorkshire 73, 85, 202; *see also* Yorkshire
Spain 68, 73; *68*
Sperrin Mountains 90
Spey Valley 89
Spurn Point 144–5
Staffordshire 20, 73, 81, 84; *22*, *113*
Stainmore 86
Stanhope 73, 86
Strathnairn 89
Strathrory 90
Strathspey 36
Strathyre 88
Sulby 92
Surrey 83
Sussex Weald 83
Sutherland 79, 89; *36*
Swinsty reservoir 22
Tain Hill 90
Tayside *109*
Thebes 70
Thorne Waste 20, 73, 85
Thornhill 88
Timble Inn 22
Timble Moor 22, 85
Tinto Hills 87
Tomatin 89
Tomintoul 81
Tweedsmuir Hills 87
Tyndrum 88
Tyne and Wear 86
Tyne Valley 86
Ukraine 280
United Kingdom 40, 42, 44, 60, 66, 123, 139, 159
United States of America 57, 135, 176, 202, 279, 281, 354
Ural Mountains 73
Wales 30, 33, 39, 56, 81; *82*
Warwickshire 169, 272, 332; *52*
Washburn Valley 22
West End 85, 22; *111*
West Kilbride 88
West Yorkshire (West Riding of Yorkshire) 85, 86, 121, 322; *see also* Yorkshire
Whitby *37*
Wicklow Mountains 90
Wigtownshire 88
Wildboarclough 21, 84; *31*, *43*, *74*
Wilhelmshaven 51
Willerby 123, 331; *236*
Winfrith Heath 20, 73
Worcestershire *52*
York 51; *see also* Jorvik
Yorkshire 22, 81, 86, 113, 139, 144, 166, 194, 203, 308, 311, 323, 339, 345; *see also* East Yorkshire, North Yorkshire, South Yorkshire, West Yorkshire, Yorkshire Dales, Yorkshire Wolds
Yorkshire Dales 322; *see also* Yorkshire
Yorkshire Wolds 184, 325; *see also* Yorkshire